Student Study Art Notebook

to accompany

Microbiology
A Human Perspective

Fourth Edition

Eugene W. Nester
University of Washington

Denise G. Anderson
University of Washington

C. Evans Roberts, Jr.
University of Washington

Nancy N. Pearsall

Martha T. Nester

Boston   Burr Ridge, IL   Dubuque, IA   Madison, WI   New York   San Francisco   St. Louis
Bangkok   Bogotá   Caracas   Kuala Lumpur   Lisbon   London   Madrid   Mexico City
Milan   Montreal   New Delhi   Santiago   Seoul   Singapore   Sydney   Taipei   Toronto

The McGraw-Hill Companies

Student Study Art Notebook to accompany
MICROBIOLOGY: A HUMAN PERSPECTIVE, FOURTH EDITION
EUGENE W. NESTER, DENISE G. ANDERSON, C. EVANS ROBERTS, JR.,
NANCY N. PEARSALL, MARTHA T. NESTER

Published by McGraw-Hill Higher Education, an imprint of The McGraw-Hill Companies, Inc., 1221 Avenue of the Americas, New York, NY 10020. Copyright © 2004 by The McGraw-Hill Companies, Inc. All rights reserved.

No part of this publication may be reproduced or distributed in any form or by any means, or stored in a database or retrieval system, without the prior written consent of The McGraw-Hill Companies, Inc., including, but not limited to, network or other electronic storage or transmission, or broadcast for distance learning.

 This book is printed on recycled, acid-free paper containing 10% postconsumer waste.

1 2 3 4 5 6 7 8 9 0 QPD/QPD 0 9 8 7 6 5 4 3

ISBN 0-07-297803-1

www.mhhe.com

DIRECTORY OF NOTEBOOK FIGURES
TO ACCOMPANY
NESTER/ANDERSON/ROBERTS/PEARSALL/NESTER
MICROBIOLOGY
A HUMAN PERSPECTIVE, 4/e

Chapter 1

Pasteur's Experiment with the Swan-Necked Flask Figure 1.2	1
"New" Infectious Diseases in Humans and Animals Since 1976 Figure 1.3	1
The Microbial World Figure 1.12	2
Sizes of Organisms and Viruses Figure 1.13	2

Chapter 2

Atom Figure 2.1	3
Covalent Bonds Figure 2.2	3
Formation of Polar Covalent Bonds in a Water Molecule Figure 2.3	4
Ionic Bond Figure 2.4	4
Weak Ionic Bonds and Molecular Recognition Figure 2.5	5
Crystal of NaCl (Salt) Resulting from Ionic Bonds Between Na^+ and Cl^- Ions Figure 2.6	5
Hydrogen Bond Formation Figure 2.7	5
Water Figure 2.8	6
Salt (NaCl) Dissolving in Water Figure 2.9	6
pH Scale Figure 2.10	7
ATP Figure 2.11	7
The Synthesis and Breakdown of Polymers Figure 2.12	7
Generalized Amino Acid Figure 2.13	7
Amino Acids Figure 2.14	8
Mirror Images (Stereoisomers) of an Amino Acid Figure 2.15	9
Peptide Bond Formation by Dehydration Synthesis Figure 2.16	9
Protein Structures Figure 2.17	10
Denaturation of a Protein Figure 2.18	10
Ribose and Deoxyribose with the Carbon Atoms Numbered Figure 2.19	11
Stereoisomers Figure 2.20	11
Structures of Three Important Polysaccharides Figure 2.21	12
A Nucleotide Figure 2.22	12
Formulas of Purines and Pyrimidines Figure 2.23	12
Joining Nucleotide Subunits Figure 2.24	13
DNA Double-Stranded Helix Figure 2.25	13
Formation of a Fat Figure 2.26	14
Fatty Acids Figure 2.27	14
Steroid Figure 2.28	14
Phospholipid and the Bilayer that Phospholipids Form in the Membrane of Cells Figure 2.29	15

Chapter 3

Refraction Figure 3.3	16
Comparison of the Principles of Light and the Electron Microscopy Figure 3.9	16
Staining Bacteria for Microscopic Observation Figure 3.13	17
Gram Stain Figure 3.14	17
Typical Shapes of Common Bacteria Figure 3.20	17
Typical Cell Groupings Figure 3.22	18
Typical Prokaryotic Cell Figure 3.23	18
The Structure of the Cytoplasmic Membrane Figure 3.24	19
Osmosis Figure 3.25	19
Proton Motive Force Figure 3.26	20
Transport Protein Figure 3.27	20
Active Transport Systems that Use Proton Motive Force Figure 3.28	21
Active Transport Systems that Use ATP Figure 3.29	21
Group Translocation Figure 3.30	21
Components and Structure of Peptidoglycan Figure 3.32	22
Gram-Positive Cell Wall Figure 3.33	23
Gram-Negative Cell Wall Figure 3.34	23
Chemical Structure of Lipopolysaccharide Figure 3.35	24
The Structure of a Flagellum in a Gram-Negative Bacterium Figure 3.39	24
Chemotaxis Figure 3.40	25
The Ribosome Figure 3.44	25
The Process of Sporulation Figure 3.47	25
Eukaryotic Cells Figure 3.48	26
Vesicle Formation and Fusion Figure 3.49	26
Endocytosis and Exocytosis Figure 3.50	26
Cytoskeleton Figure 3.51	27
Flagella Figure 3.52	27
Nucleus Figure 3.53	28
Chromatin Figure 3.54	28
Mitochondria Figure 3.55	29
Chloroplasts Figure 3.56	29

Endoplasmic Reticulum   Figure 3.57   30
Golgi Apparatus   Figure 3.58   30

Chapter 4
The Streak-Plate Method   Figure 4.2   31
Binary Fission   Figure 4.3   32
Temperature Requirements for Growth
  Figure 4.4   32
Effects of Solute Concentration on Cells
  Figure 4.5   33
Anaerobe Jar   Figure 4.8   33
Enrichment Culture   Figure 4.10   34
A Counting Chamber   Figure 4.11   34
A Coulter Counter   Figure 4.12   35
Plate Counts   Figure 4.13   36
The Most Probable Number (MPN) Method
  Figure 4.15   37
Measuring Turbidity with a
  Spectrophotometer   Figure 4.16   37
Growth Curve   Figure 4.17   38
Primary and Secondary Metabolite
  Production   Figure 4.18   38
Dynamic Population Changes in the Phase
  of Prolonged Decline   Figure 4.19   38

Chapter 5
The Relationship Between the Numbers of
  Initial Microorganisms and the Time It
  Takes to Kill Them   Figure 5.2   39
Steam-Jacketed Autoclave   Figure 5.3   39
Steps in the Commercial Canning of Foods
  Figure 5.5   40
Sites of Action of Germicidal Chemical
  Figure 5.6   41
Filtration   Figure 5.7   41
The Electromagnetic Spectrum   Figure 5.8   42

Chapter 6
The Relationship Between Catabolism and
  Anabolism   Figure 6.1   43
Most Chemoorganotrophs Depend on the
  Radiant Energy Harvested by
  Photosynthetic Organisms   Figure 6.3   44
Energetics of Chemical Reactions
  Figure 6.4   45
Metabolic Pathways   Figure 6.5   45
The Role of Enzymes   Figure 6.6   46
ATP   Figure 6.7   46
Oxidation-Reduction Reactions
  Figure 6.8   46
Scheme of Metabolism   Figure 6.9   47
Mechanism of Enzyme Action
  Figure 6.10   48
Enzymes May Act in Conjunction with a
  Cofactor   Figure 6.11   48
Environmental Factors that Influence
  Enzyme Activity   Figure 6.12   48
Regulation of Allosteric Enzymes
  Figure 6.13   49
Competitive Inhibition of Enzymes
  Figure 6.14   49
Glycolysis   Figute 6.15   50
The Transition Step and the Tricarboxylic
  Acid Cycle   Figure 6.16   51
Electron Transport   Figure 6.17   52
The Electron Transport Chain of Mitochondria
  Figure 6.18   52
The Electron Transport Chain of *E. coli*
  Growing Aerobically in a Glucose-
  Containing Medium   Figure 6.19   53
Anaerobic Respiration   Figure 6.20   53
Maximum Theoretical Energy Yield from
  Aerobic Respiration in a Prokaryotic Cell
  Figure 6.21   54
Fermentation Pathways Use Pyruvate or a
  Derivative As a Terminal Electron
  Acceptor   Figure 6.22   55
End Products of Fermentation Pathways
  Figure 6.23   55
Catabolism of Organic Compounds Other
  than Glucose   Figure 6.24   56
Relative Energy Gain of Different Types of
  Metabolism   Figure 6.25   57
Photosystem   Figure 6.26   57
The Tandem Photosystems of Cyanobacteria
  and Chloroplasts   Figure 6.27   58
The Calvin Cycle   Figure 6.28   59
The Use of Precursor Metabolites in
  Biosynthesis   Figure 6.29   60
Glutamate   Figure 6.30   61
Synthesis of Aromatic Amino Acids
  Figure 6.31   61
Source of the Carbons and Nitrogen Atoms
  in Purine Rings   Figure 6.32   61

Chapter 7
Overview of Replication, Transcription, and
  Translation   Figure 7.1   62
Diagrammatic Representations of the
  Structure of DNA   Figure 7.2   62
The Double Helix of DNA   Figure 7.3   63
Replication of Chromosomal DNA of
  Prokaryotes   Figure 7.4   63
The Process of DNA Synthesis   Figure 7.5   64
The Replication Fork   Figure 7.6   64
RNA Is Transcribed from a DNA Template
  Figure 7.7   65
Promoters Direct Transcription   Figure 7.8   65
The Process of RNA Synthesis   Figure 7.9   66
The Genetic Code   Figure 7.10   67
Reading Frames   Figure 7.11   67
The Structure of the 70S Ribosome
  Figure 7.12   67
The Structure of Transfer RNA (tRNA)
  Figure 7.13   68
In Prokaryotes, Translation Begins As the
  mRNA Molecule Is Still Being
  Synthesized   Figure 7.14   68

The Process of Translation Figure 7.15	69
Splicing of Eukaryotic RNA Figure 7.16	70
Transcriptional Regulation by Repressors Figure 7.18	70
Transcriptional Regulation by Activators Figure 7.19	71
Regulation of the *lac* Operon Figure 7.20	71
Diauxic Growth Curve of *E. coli* Growing in a Medium Containing Glucose and Lactose Figure 7.21	72

Chapter 8

Base Substitution Figure 8.1	73
Frameshift Mutation As a Result of Base Addition Figure 8.2	74
Common Base Analogs and the Normal Bases They Replace in DNA Figure 8.4	74
Thymine Dimer Formation Figure 8.5	75
Mismatch Repair Figure 8.6	75
Repair of Thymine Dimers Figure 8.7	76
Direct Selection of Mutants Figure 8.8	76
Indirect Selection of Mutants by Replica Plating Figure 8.9	77
Penicillin Enrichment of Mutants Figure 8.10	77
Isolation of Temperature-Sensitive (Conditional) Mutants Figure 8.11	78
Ames Test to Screen for Mutagens Figure 8.12	78
General Experimental Approach for Detecting Gene Transfer in Bacteria Figure 8.13	79
DNA-Mediated Transformation Figure 8.14	80
Electroporation Figure 8.15	80
Transduction (Generalized) Figure 8.16	81
Conjugation—Transfer of the F Plasmid Figure 8.18	82
Hfr Formation Figure 8.19	83
Conjugation—Transfer of Chromosomal DNA Figure 8.20	83
Formation of F′ Plasmid Figure 8.21	84
Two Regions of an R Plasmid Figure 8.22	84
The Movement of a Transposon Throughout a Bacterial Population Figure 8.23	85
Transposable Elements Figure 8.24	85
Action of Restriction Endonucleases Figure 8.25	86
Restriction and Modification of Entering DNA Figure 8.26	86

Chapter 9

DNA Must Replicate in a Cell in Order to be Maintained in a Population of Cells Figure 9.1	87
The Steps of a Cloning Experiment Figure 9.2	87
Cloning into a High-Copy-Number Plasmid Figure 9.3	87
The Function of a Reporter Gene Figure 9.4	88
Nucleic Acid Hybridization Can Be Used to Locate Homologous Sequences Figure 9.6	89
The Steps of a Colony Blot Figure 9.7	90
The Steps of a Southern Blot Figure 9.8	91
Restriction Fragment Length Polymorphism (RFLP) Figure 9.9	92
PCR Amplifies Selected Sequences Figure 9.11	92
A DNA Library Figure 9.12	93
Making cDNA from Eukaryotic mRNA Figure 9.13	94
Cohesive ends Figure 9.14	95
Typical Properties of an Ideal Vector Figure 9.15	96
The Function of the *lacZ'* Gene in a Vector Figure 9.16	96
Chain Termination by a Dideoxynucleotide Figure 9.19	97
Dideoxy Chain Termination Method for Determining the Nucleotide Sequence of DNA Figure 9.20	97
Steps of a Single Cycle of PCR Figure 9.23	98
The Final PCR Product Is a Fragment of Discrete Size Figure 9.24	99

Chapter 10

The Three-Domain System of Classification Figure 10.1	100
An Example of a Dichotomous Key Leading to the Identification of *E. coli* Figure 10.5	101
Chromatogram of the Fatty Acid Profiles Figure 10.7	101
Nucleic Acid Probes to Detect Specific DNA Sequences Figure 10.8	102
Ribosomal RNA Figure 10.9	102
Phage typing Figure 10.12	103
Numerical Taxonomy Figure 10.15	103
A DNA Melting Curve Figure 10.16	104
Using DNA Hybridization to Assess Relatedness Figure 10.17	104
A Phylogenetic Tree Figure 10.18	105

Chapter 11

Schematic Drawing of a Member of the Family *Enterobacteriaceae* Figure 11.15	106
Caulobacter Figure 11.23	106
Hyphomicrobium Figure 11.24	107
Bdellovibrio Figure 11.25	107

Chapter 12

A Phylogeny of the Eukaryotes Based on Ribosomal RNA Sequence Comparison Figure 12.1	108

Phylogeny of Algae   Figure 12.3   109
Binary Fission Is an Asexual Reproduction Process in Which a Single Cell Divides into Two Independent Daughter Cells   Figure 12.5   110
A Phylogenetic Scheme of Eukaryotes Based on rRNA Sequence Comparisons   Figure 12.7   110
Various Forms of Asexual Reproduction in Protozoa   Figure 12.10   111
A Phylogenetic Scheme of Eukaryotes Based on rRNA Sequence Comparisons   Figure 12.11   111
Budding in Yeast   Figure 12.14   112
Formation of Hyphae and Mycelium   Figure 12.15   112
A Meadow Mushroom, *Agaricus campestri*   Figure 12.17   113
Lichens   Figure 12.18   113
A Phylogenetic Scheme of Eukaryotes Based on rRNA Sequence Comparisons   Figure 12.19   114
Slime Molds   Figure 12.20   114
Internal Anatomy of a Mosquito   Figure 12.21   115
Sarcoptes scabiei (Scabies Mite)   Figure 12.23   115
Life Cycle of the Pork Tapeworm, *Taenia solium,* Acquired by Eating Inadequately Cooked Pork   Figure 12.24   116

Chapter 13
Common Shapes of Viruses   Figure 13.1   117
Two Different Types of Virions   Figure 13.2   117
Virion Size   Figure 13.3   118
Major Types of Relationships Between Viruses and the Host Cells They Infect   Figure 13.4   118
Steps in the Replication of T4 Phage in *E. coli*   Figure 13.5   119
Lambda Phage (λ) Replication Cycle   Figure 13.6   120
Reversible Insertion and Excision of Lambda (λ) Phage   Figure 13.7   121
Replication of a Filamentous Phage   Figure 13.9   122
Macromolecule Synthesis in Filamentous Phage Replication   Figure 13.10   122
Specialized Transduction by Temperate Phage   Figure 13.11   123
Restriction-Modification System   Figure 13.13   124

Chapter 14
Preparation of Primary Cell Culture   Figure 14.2   125
Hemagglutination   Figure 14.6   126
Time Course of Appearance of Symptoms of Measles and the Measles Virions   Figure 14.7   126
Entry of Enveloped Animal Viruses into Host Cells   Figure 14.8   127
Strategies of Transcription Employed by Different Viruses   Figure 14.9   128
Tobacco Mosaic Virus Assembly   Figure 14.10   128
Mechanism for Releasing Enveloped Virions   Figure 14.11   129
Time Course of Appearance of Disease Symptoms and Infectious Virions in Various Kinds of Viral Infections   Figure 14.12   130
Infection Cycle of Herpes Simplex Virus, HSV-1   Figure 14.13   131
Replication Cycle of a Retrovirus   Figure 14.14   132
Various Effects of Animal Viruses on the Cells They Infect   Figure 14.15   133
Phenotypic Mixing   Figure 14.16   134
Phenotypic Mixing of Two Retroviruses   Figure 14.17   134
Genetic Reassortment   Figure 14.18   135
Proposed Mechanism by Which Prions Replicate   Figure 14.22   135

Chapter 15
Anatomical Barriers   Figure 15.1   136
Epithelial Barriers   Figure 15.2   136
First-Line Defense Mechanisms in Humans   Figure 15.3   137
Blood and Lymphoid Cells   Figure 15.4   138
Mononuclear Phagocyte System   Figure 15.5   139
Toll-Like Receptors   Figure 15.6   139
Complement System   Figure 15.7   140
Membrane Attack Complex of Complement (MAC)   Figure 15.8   141
Phagocytosis and Intracellular Destruction of Phagocytized Material   Figure 15.9   141
The Inflammatory Process   Figure 15.10   142
Mechanism of the Antiviral Activity of Interferons   Figure 15.11   143

Chapter 16
Overview of Humoral and Cellular Immunity   Figure 16.1   144
Anatomy of the Lymphoid System   Figure 16.2   145
Antibodies and Antigen Epitopes on a Bacterial Cell   Figure 16.3   146
Basic Structure of an Antibody Molecule   Figure 16.4   146
Model of an IgG Molecule   Figure 16.5   147
Protective Outcomes of Antibody-Antigen Binding   Figure 16.6   148
Immunoglobulin G Levels in the Fetus and Infant   Figure 16.7   149
Clonal Selection and Expansion During the Antibody Response   Figure 16.8   149

Antigen Presentation by a B Cell
  Figure 16.9 — 150
The Primary and Secondary Responses to
  Antigen   Figure 16.11 — 150
Affinity Maturation   Figure 16.12 — 151
Class Switching   Figure 16.13 — 152
T-Independent Antigens   Figure 16.14 — 152
Two T-Cell Receptors Composed of Alpha
  and Beta Polypeptide Chains
  Figure 16.15 — 153
MHC Molecules   Figure 16.16 — 153
Antigen Recognition by T Cells   Figure 16.17 — 153
Consequences of Antigen Recognition by
  Effector T-Cytotoxic Cells   Figure 16.18 — 154
Antigen Presentation by a Macrophage to
  an Effector T-Helper Cell   Figure 16.19 — 155
Activation of T Cells by Dendritic Cells
  Expressing Co-Stimulatory Molecules
  Figure 16.20 — 156
Summary of the Adaptive Immune
  Response   Figure 16.21 — 157
Antibody Diversity   Figure 16.22 — 158

Chapter 17
Acquired Immunity   Figure 17.1 — 159
Quantitation of Immunologic Tests
  Figure 17.2 — 159
Antigen-Antibody Precipitation Reactions
  Figure 17.3 — 160
Radial Immunodiffusion Is Used to Measure
  the Concentration of Antigen in a Sample
  Figure 17.4 — 161
Double Immunodiffusion   Figure 17.5 — 162
Immunoelectrophoresis Permits
  Identification of Antigens in a Mixture
  Figure 17.6 — 163
Fluorescent Antibody Tests   Figure 17.8 — 164
Enzyme-Linked Immunosorbent Assay
  (ELISA)   Figure 17.9 — 165
Complement Fixation Test Procedure
  Figure 17.12 — 166

Chapter 18
Mechanisms of Type I Hypersensitivity:
  Immediate IgE-Mediated   Figure 18.1 — 167
Immunotherapy for IgE Allergies
  Figure 18.3 — 168
Hemolytic Disease of the Newborn
  Figure 18.4 — 169
Type III Hypersensitivity: Immune Complex-
  Mediated   Figure 18.5 — 170
Poison Oak Dermatitis Is an Example of
  Type IV Hypersensitivity: Delayed Cell-
  Mediated   Figure 18.7 — 171

Chapter 19
Normal Flora   Figure 19.1 — 172
The Course of Infectious Diseases
  Figure 19.2 — 172
Pili Attachment to Host Cell   Figure 19.3 — 173
Type III Secretion Systems   Figure 19.4 — 173
Antigen-Sampling Processes Provide a
  Mechanism for Invasion   Figure 19.6 — 174
Avoiding the Alternative Pathway of
  Complement Activation   Figure 19.8 — 174
Avoiding Destruction by Phagocytes
  Figure 19.9 — 175
Foiling Opsonization by Antibodies
  Figure 19.11 — 175
The Action of A-B Exotoxins   Figure 19.12 — 176
Superantigens   Figure 19.13 — 176
Endotoxin   Figure 19.14 — 177
Cytomegalovirus-Immune Cell Interactions
  Figure 19.15 — 178

Chapter 20
Spread of Pathogens   Figure 20.1 — 179
Incidence of Influenza, and Endemic
  Disease that Can Be Epidemic
  Figure 20.2 — 179
Incidence of Tetanus by Age Group
  Figure 20.4 — 180
Comparison of Propagated Versus
  Common Source Epidemics   Figure 20.5 — 180
Seasonal Occurrence of Respiratory
  Infections Caused by Respiratory
  Syncytial Virus   Figure 20.6 — 181
Seasonal Occurrence of Gastrointestinal
  Diseases   Figure 20.7 — 181
World Map of Emerging Diseases
  Figure 20.10 — 182
Relative Frequency of Different Types of
  Nosocomial Infections   Figure 20.11 — 182

Chapter 21
Family Tree of Penicillins   Figure 21.1 — 183
Targets of Antibacterial Medications
  Figure 21.2 — 183
Antibacterial Medications that Interfere
  with Cell Wall Biosynthesis
  Figure 21.3 — 184
The β-Lactam Ring of Penicillins and
  Cephalosporins   Figure 21.4 — 184
Chemical Structures and Properties of
  Representative Members of the
  Penicillin Family   Figure 21.6 — 185
Antibacterial Medications that Inhibit
  Prokaryotic Protein Synthesis
  Figure 21.7 — 186
Inhibitors of the Folate Pathway
  Figure 21.8 — 186
Determining the Minimum Inhibitory
  Concentration (MIC) of an Antimicrobial
  Drug   Figure 21.9 — 187
The Selective Advantage of Drug
  Resistance   Figure 21.13 — 188
Common Mechanisms of Antimicrobial Drug
  Resistance   Figure 21.14 — 189

The Acquisition of Antimicrobial Resistance
  Figure 21.15 — 189
Targets of Antiviral Drugs   Figure 21.16 — 190
Targets of Antifungal Drugs   Figure 21.17 — 190

Chapter 22
Microscopic Anatomy of the Skin
  Figure 22.1 — 191
Pathogenesis of a Boil (Furuncle)
  Figure 22.3 — 192
Pathogenesis of Post-Streptococcal Acute
  Glomerulonephritis   Figure 22.7 — 193
Total Reported Cases of Rocky Mountain
  Spotted Fever by State and Region,
  1994–1998   Figure 22.10 — 194
Average Number of Reported Cases of
  Lyme Disease per Year 1990–1999
  Figure 22.14 — 194
Life Cycle of the Black-Legged Tick, *Ixodes
  scapularis*, the Principal Vector of
  Borrelia burgdorferi, Cause of Lyme
  Disease   Figure 22.16 — 195
Reported Incidence of Measles in Different
  Regions of the World, 1990–1998
  Figure 22.22 — 196
Reported Cases of German Measles
  (Rubella), United States, 1967–2001
  Figure 22.24 — 196

Chapter 23
Anatomy and Infections of the Respiratory
  System   Figure 23.1 — 197
Components of the Cell Envelope of
  Streptococcus pyogenes   Figure 23.3 — 198
Rheumatic Heart Disease   Figure 23.4 — 199
Mode of Action of Diphtheria Toxin
  Figure 23.6 — 200
Otitis Media   Figure 23.7 — 201
Mode of Action of Pertussis Toxin
  Figure 23.14 — 202
Number of Reported Pertussis Cases, by
  Year, United States, 1922–2000
  Figure 23.15 — 203
Incidence of Tuberculosis, United States,
  1978–2001   Figure 23.16 — 203
Diagrammatic Representation of Influenza
  Virus   Figure 23.21 — 204
Influenza Virus: Antigenic Drift and Antigenic
  Shift   Figure 23.22 — 204
Hantavirus Pulmonary Syndrome Cases,
  United States, as of January 30, 2002
  Figure 23.23 — 205
Area of Distribution of *Coccidioides Immitis*
  Figure 23.25 — 205
Geographic Distribution of *Histoplasma
  capsulatum* in the United States as
  Revealed by Positive Skin Tests
  Figure 23.27 — 205

Chapter 24
The Alimentary System   Figure 24.1 — 206
Structure of a Tooth and Its Surrounding
  Tissues   Figure 24.2 — 207
Increase in Acidity in Cariogenic Dental
  Plaque After Rinsing the Mouth with a
  Glucose Solution   Figure 24.4 — 207
Helicobacter pylori   Figure 24.7 — 208
Gastric Ulcer Formation Associated with
  Helicobacter pylori Infection
  Figure 24.8 — 208
Herpes Simplex Labialis, Also Know As
  Cold Sores or Fever Blisters
  Figure 24.9 — 209
Reported Cases of Mumps, United States,
  1973–2001   Figure 24.11 — 209
Mode of Action of Cholera Toxin
  Figure 24.13 — 210
Pathogenesis of Shigellosis   Figure 24.14 — 211
Reported Cases of Hepatitis A, United
  States, 2001   Figure 24.18 — 211
Hepatitis B Virus Components Found in the
  Blood of Infected Individuals
  Figure 24.19 — 212
Replication of Hepatitis B Virus
  Figure 24.20 — 213
Incidence of Viral Hepatitis in the United
  States   Figure 24.21 — 214
Life Cycle of *Entamoeba histolytica*
  Figure 24.24 — 214

Chapter 25
Anatomy of the Urinary System   Figure 25.1 — 215
Anatomy of the Genital System
  Figure 25.2 — 216
Staphylococcal Toxic Shock, United States,
  1979–1996   Figure 25.6 — 216
The Possible Risk of Acquiring a Sexually
  Transmitted Disease in Two Individuals
  Contemplating Unprotected Sexual
  Intercourse   Figure 25.7 — 217
The AIDS Pandemic Continues Without
  Letup   Figure 25.20 — 218
Trichomonas vaginalis, a Common Sexually
  Transmitted Cause of Vaginitis
  Figure 25.21 — 218

Chapter 26
The Central Nervous System   Figure 26.1 — 219
Cerebrospinal Fluid   Figure 26.2 — 220
Rate of Serious *Haemophilus Influenzae*
  Disease per 100,000 Children Less than
  Age Five, United States, 1987 through
  1998   Figure 26.3 — 221
The Five Leading Causes of Meningitis in
  the United States, 1995   Figure 26.5 — 221
Meningococcal Disease in the United
  States, 1935 to 1998   Figure 26.8 — 222

LaCrosse Encephalitis Virus, Natural Cycles
Figure 26.13 — **222**
Distribution of Encephalitis-Causing Arboviruses Figure 26.14 — **223**
Incidence of Poliomyelitis in the United States, 1951 to 1998 Figure 26.18 — **223**

Chapter 27
The Process of Wound Repair Figure 27.1 — **224**
Abscess Formation Figure 27.2 — **225**
Tetanus Figure 27.9 — **226**
Inhibitory Neuron Function Figure 27.10 — **226**
Average Annual Incidence of Tetanus for Different Age Groups, United States, 1995 to 1997 Figure 27.11 — **227**

Chapter 28
The Blood and Lymphatic Systems Figure 28.1 — **228**
Events in Gram-Negative Septicemia Figure 28.3 — **229**
Reported Cases of Tularemia, United States, 1990–2000 Figure 28.5 — **230**
Pathogenesis of Infectious Mononucleosis Figure 28.8 — **230**
Distribution of Yellow Fever Figure 28.10 — **231**
Life Cycle of *Plasmodium vivax* Figure 28.11 — **231**
Distribution of Malaria in 1996 Figure 28.12 — **232**

Chapter 29
The HIV/AIDS Epidemic Two Decades After the Onset Figure 29.1 — **233**
Human Immunodeficiency Virus, Type 1 (HIV-1) Figure 29.3 — **234**
Some of HIV's Cellular Targets Figure 29.4 — **235**
Attachment and Entry of HIV into a Host Cell, Schematic Representation Figure 29.5 — **236**
The Steps in HIV Replication Figure 29.6 — **237**
Diagram of the SU Glycoprotein Figure 29.7 — **238**
Natural History of HIV Disease Figure 29.8 — **238**
In the United States, a Steadily Rising Percentage of AIDS Cases Occur in Women Figure 29.9 — **239**
Estimated Deaths Due to AIDS, United States, 1993 to 1998 Figure 29.10 — **239**
Mode of Action of Zidovudine (AZT) Figure 29.11 — **240**
Toxoplasma gondii Figure 29.14 — **241**

Chapter 30
Relationship Between Producers, Consumers, and Decomposers in an Ecosystem Figure 30.2 — **242**
Competition Between Two Bacteria Figure 30.3 — **243**
Growth of Microbial Populations in Unpasteurized Raw Milk at Room Temperature Figure 30.4 — **243**
Thermal Stratification of a Lake Figure 30.7 — **244**
Texture of Soil Figure 30.8 — **244**
Carbon Cycle Figure 30.9 — **245**
Nitrogen Cycle Figure 30.11 — **246**
Sulfur Cycle Figure 30.12 — **246**
Hydrothermal Vent Community Figure 30.13 — **247**
Symbiotic Nitrogen Fixation Figure 30.14 — **247**

Chapter 31
Municipal Sewage Treatment Figure 31.1 — **248**
Trickling Filter Figure 31.2 — **249**
Artificial Wetland Figure 31.3 — **250**
Septic Tank Figure 31.4 — **251**
Groundwater Figure 31.5 — **252**
Steps in the Treatment of Metropolitan Water Supplies Figure 31.6 — **253**
Methods Used for Testing Water Figure 31.7 — **254**
Industrial and Backyard Composting Figure 31.9 — **255**
Comparison of the Rates of Disappearance of Two Structurally Related Herbicides, 2,4-D, and 2,4,5-T Figure 31.10 — **256**

Chapter 32
Commercial Production of Wine Figure 32.4 — **257**
Commercial Production of Beer Figure 32.5 — **258**

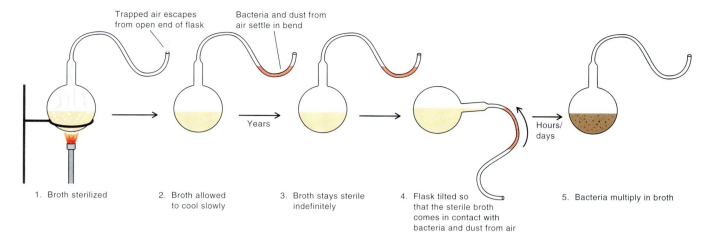

Pasteur's Experiment with the Swan-Necked Flask
Figure 1.2

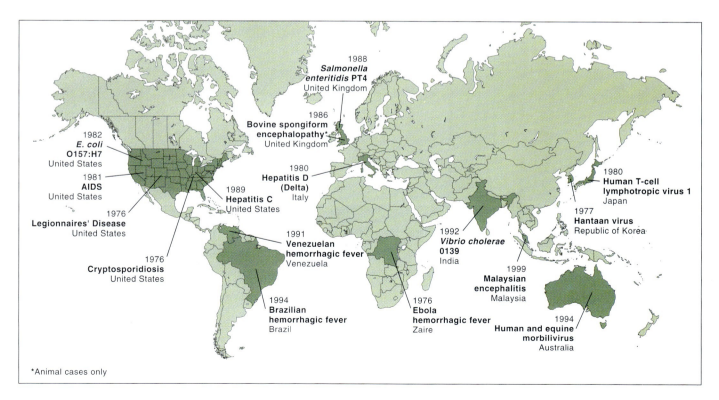

"New" Infectious Diseases in Humans and Animals Since 1976
Figure 1.3

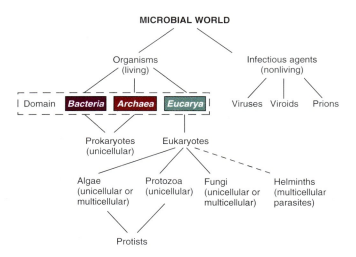

The Microbial World
Figure 1.12

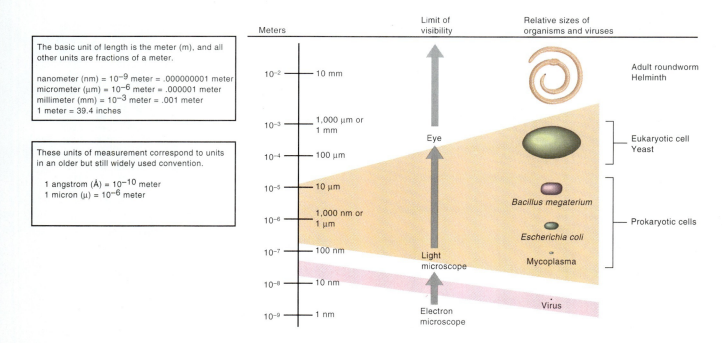

Sizes of Organisms and Viruses
Figure 1.13

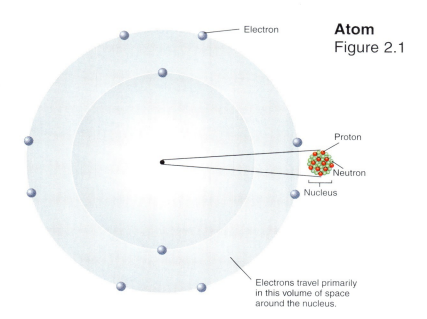

Atom
Figure 2.1

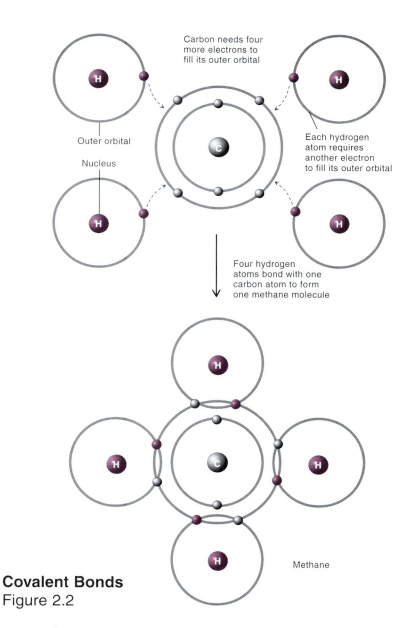

Covalent Bonds
Figure 2.2

Formation of Polar Covalent Bonds in a Water Molecule
Figure 2.3

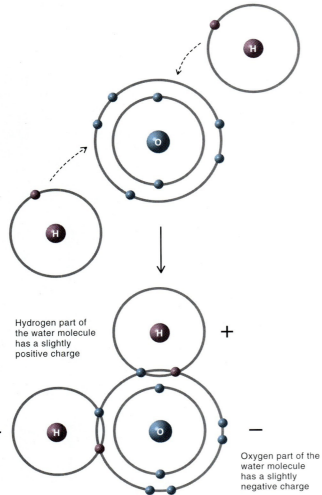

Hydrogen part of the water molecule has a slightly positive charge

Oxygen part of the water molecule has a slightly negative charge

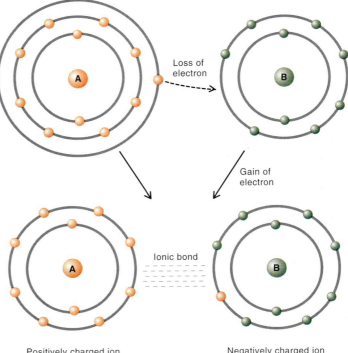

Ionic Bond
Figure 2.4

Positively charged ion

Negatively charged ion

Weak Ionic Bonds and Molecular Recognition
Figure 2.5

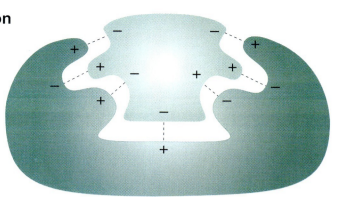

Crystal of NaCl (Salt) Resulting from Ionic Bonds Between Na⁺ and Cl⁻ Ions
Figure 2.6

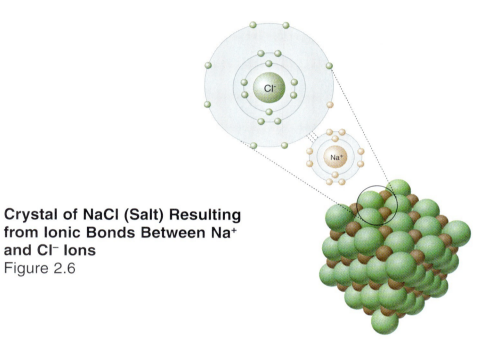

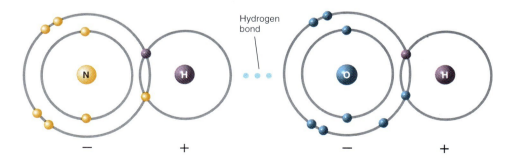

Hydrogen Bond Formation
Figure 2.7

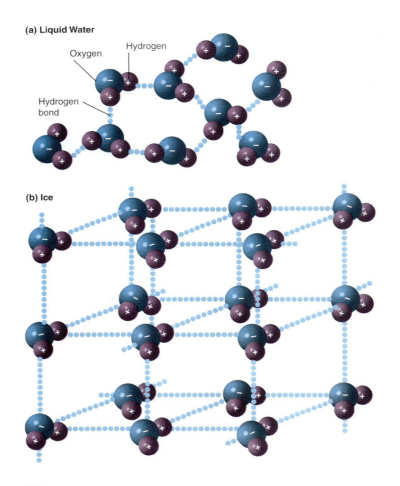

Water
Figure 2.8

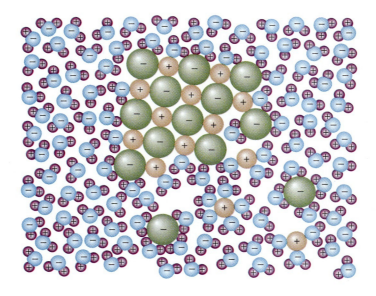

Salt (NaCl) Dissolving in Water
Figure 2.9

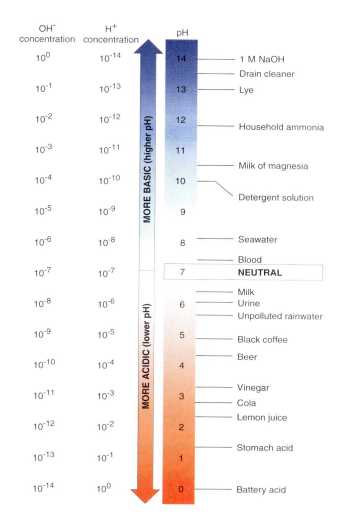

pH Scale
Figure 2.10

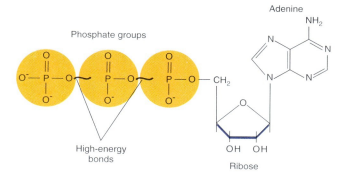

ATP
Figure 2.11

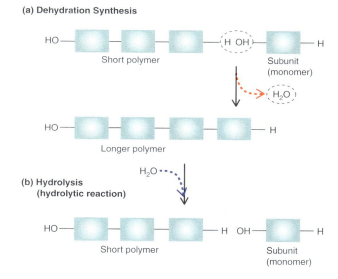

The Synthesis and Breakdown of Polymers
Figure 2.12

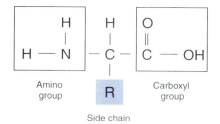

Generalized Amino Acid
Figure 2.13

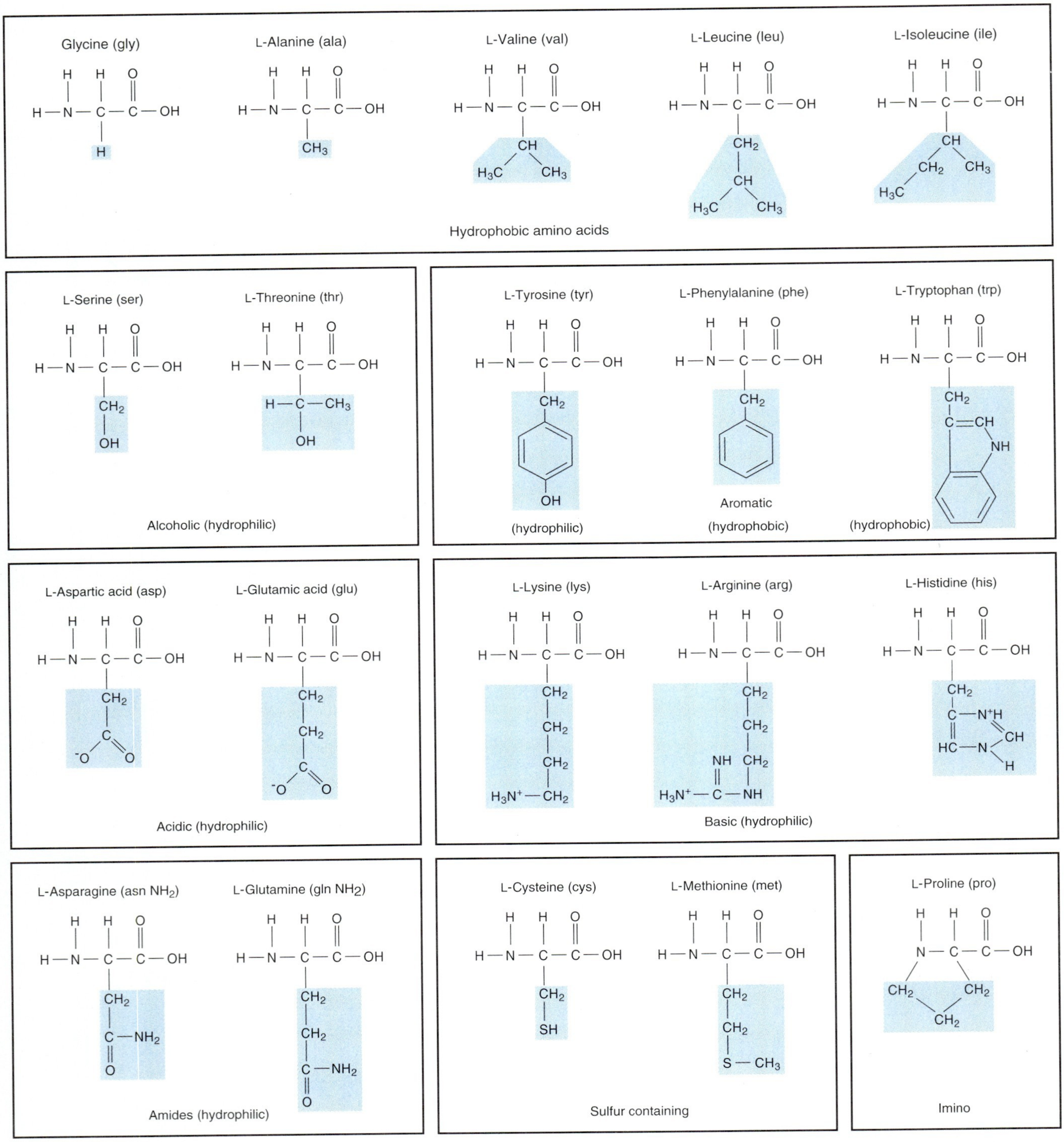

Amino Acids
Figure 2.14

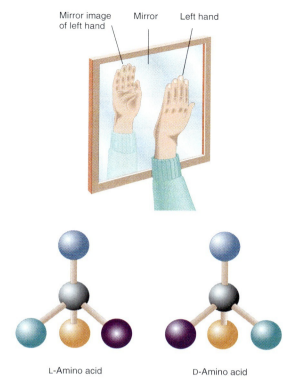

Mirror Images (Stereoisomers) of an Amino Acid
Figure 2.15

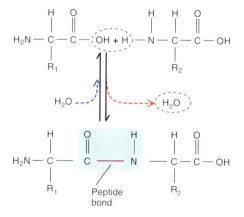

Peptide Bond Formation by Dehydration Synthesis
Figure 2.16

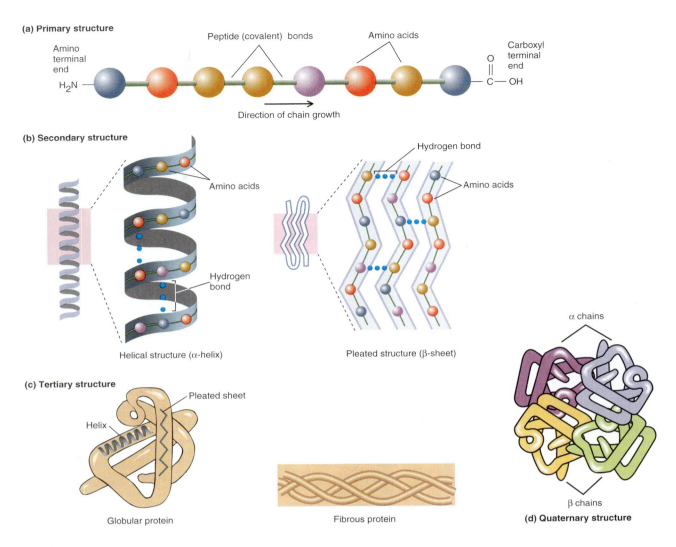

Protein Structures
Figure 2.17

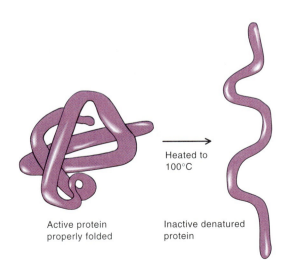

Denaturation of a Protein
Figure 2.18

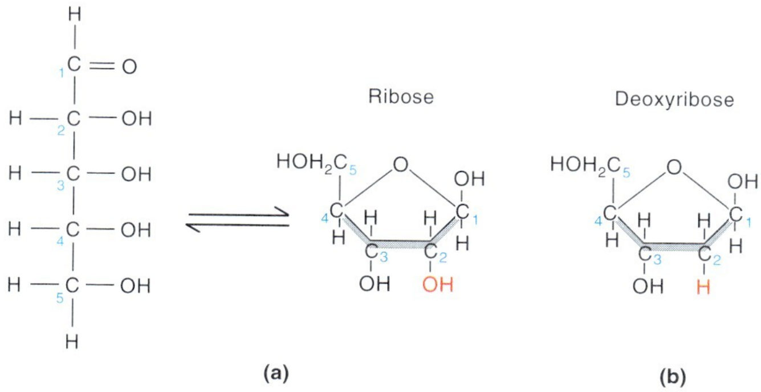

Ribose and Deoxyribose with the Carbon Atoms Numbered
Figure 2.19

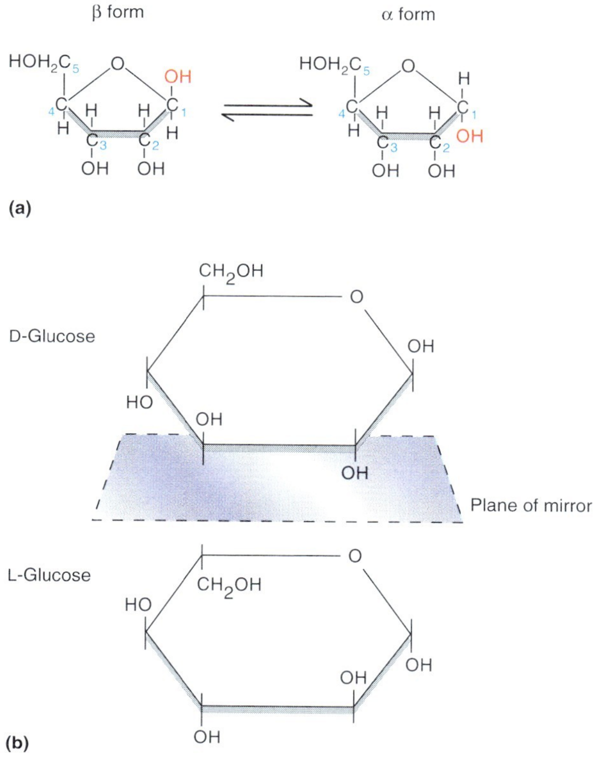

Stereoisomers
Figure 2.20

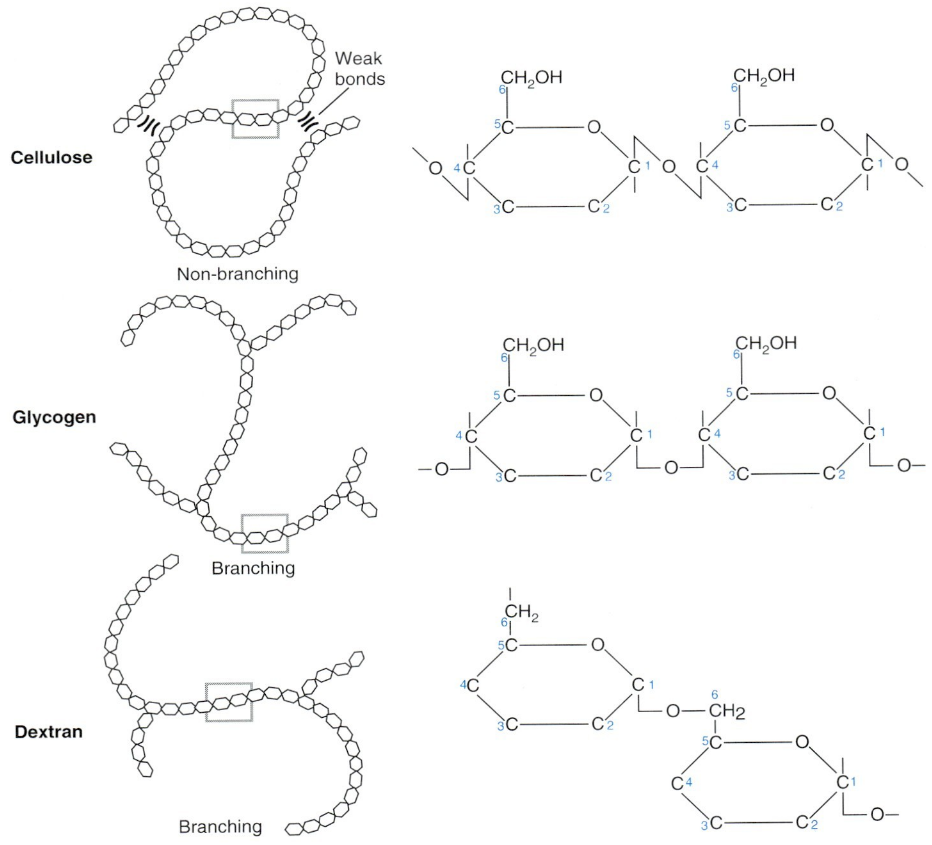

Structures of Three Important Polysaccharides
Figure 2.21

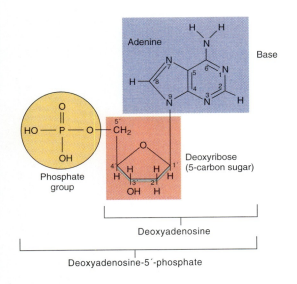

A Nucleotide
Figure 2.22

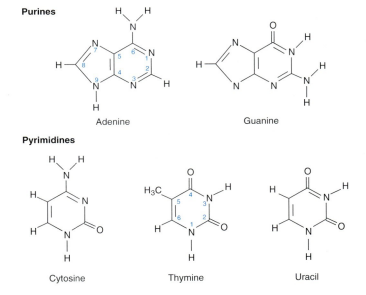

Formulas of Purines and Pyrimidines
Figure 2.23

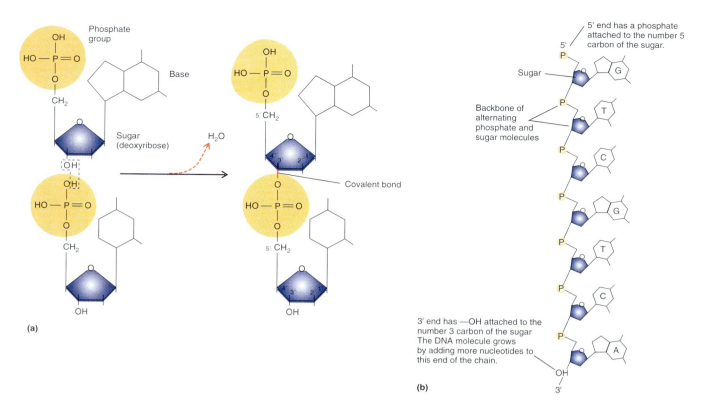

Joining Nucleotide Subunits
Figure 2.24

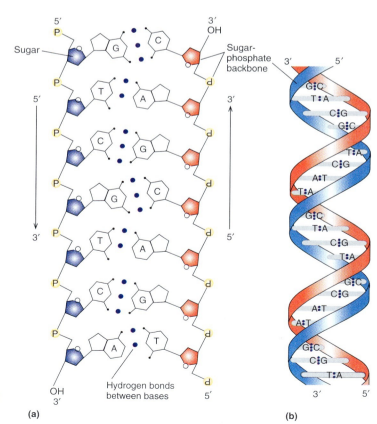

DNA Double-Stranded Helix
Figure 2.25

13

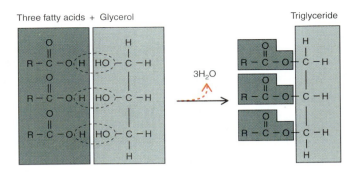

Formation of a Fat
Figure 2.26

(a) Palmitic acid

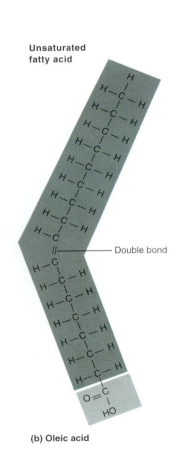

(b) Oleic acid

Fatty Acids
Figure 2.27

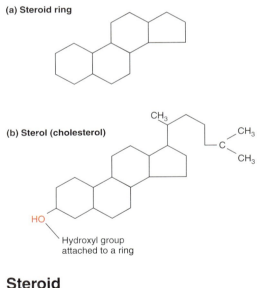

Steroid
Figure 2.28

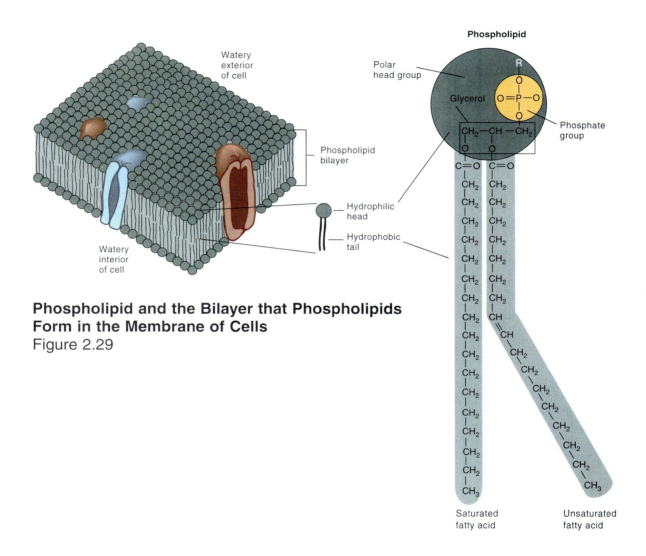

Phospholipid and the Bilayer that Phospholipids Form in the Membrane of Cells
Figure 2.29

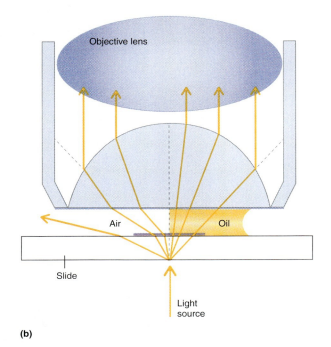

Refraction
Figure 3.3

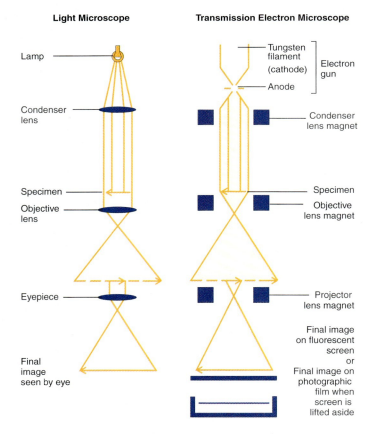

Comparison of the Principles of Light and the Electron Microscopy
Figure 3.9

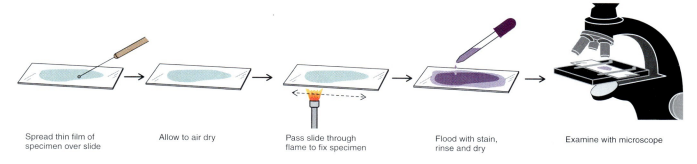

Staining Bacteria for Microscopic Observation
Figure 3.13

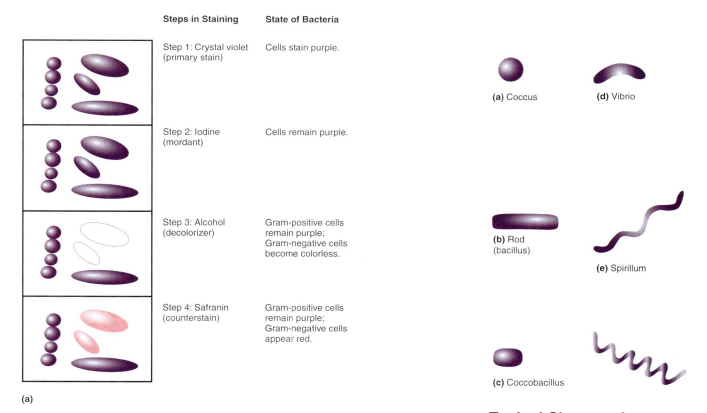

Gram Stain
Figure 3.14

Typical Shapes of Common Bacteria
Figure 3.20

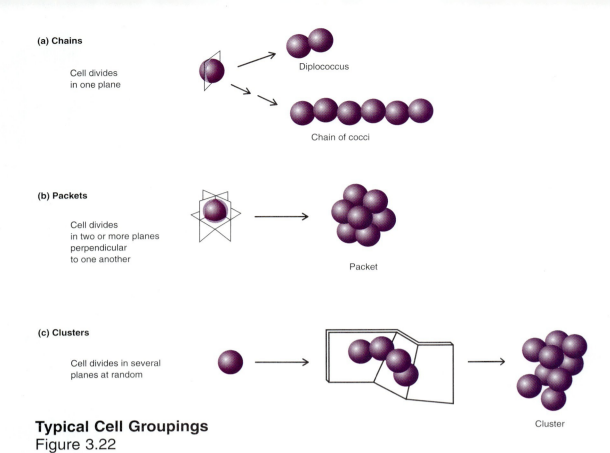

Typical Cell Groupings
Figure 3.22

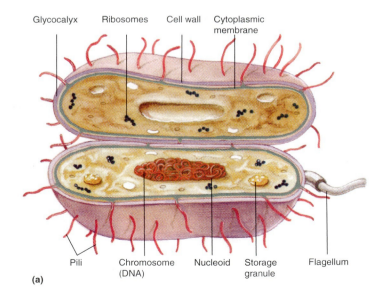

Typical Prokaryotic Cell
Figure 3.23

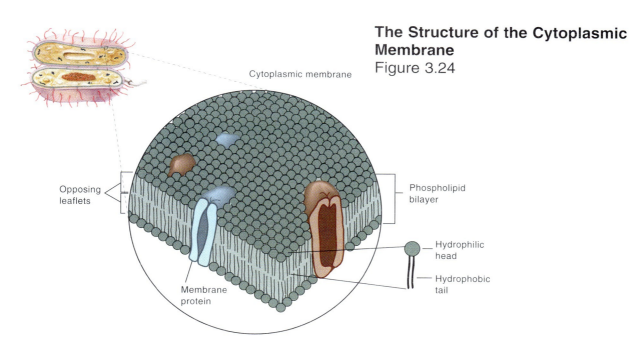

The Structure of the Cytoplasmic Membrane
Figure 3.24

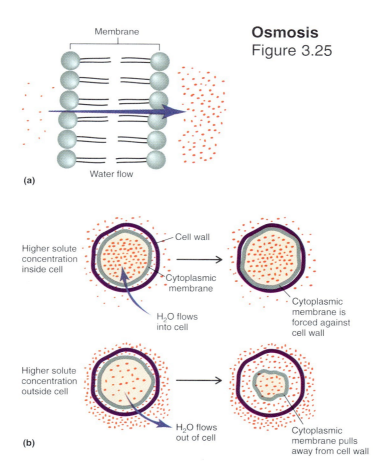

Osmosis
Figure 3.25

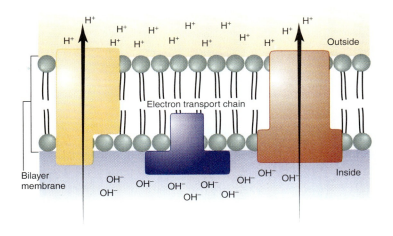

Proton Motive Force
Figure 3.26

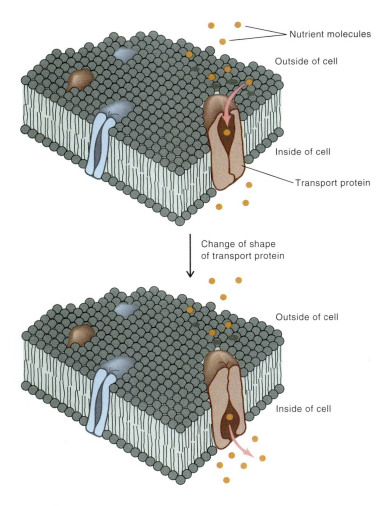

Transport Protein
Figure 3.27

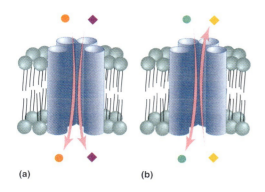

Active Transport Systems that Use Proton Motive Force
Figure 3.28

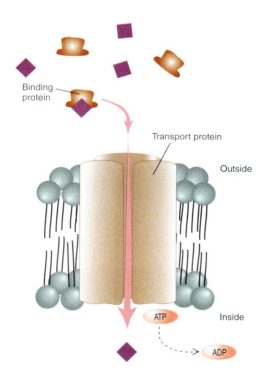

Active Transport Systems that Use ATP
Figure 3.29

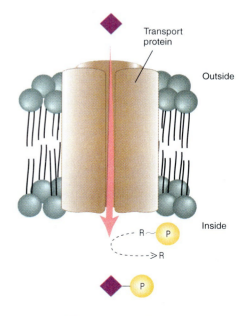

Group Translocation
Figure 3.30

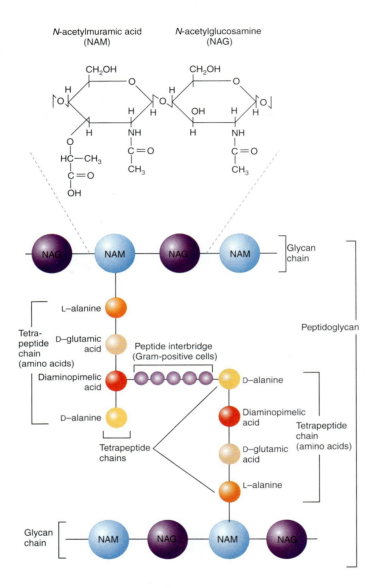

Components and Structure of Peptidoglycan
Figure 3.32

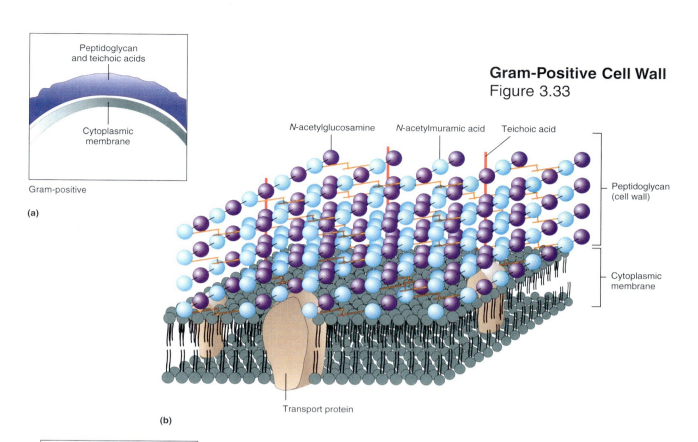

Gram-Positive Cell Wall
Figure 3.33

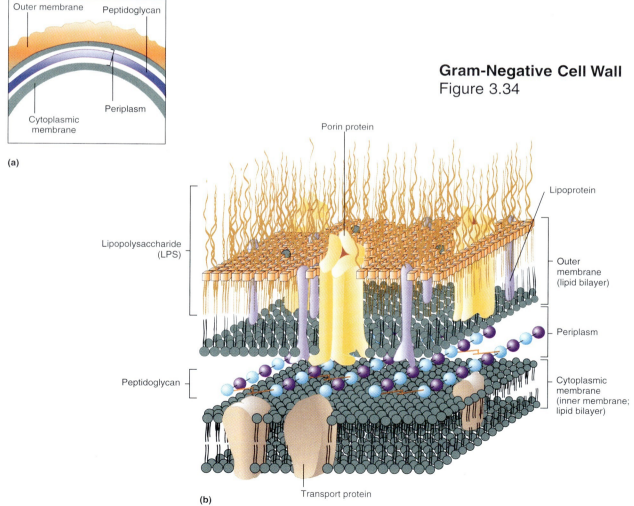

Gram-Negative Cell Wall
Figure 3.34

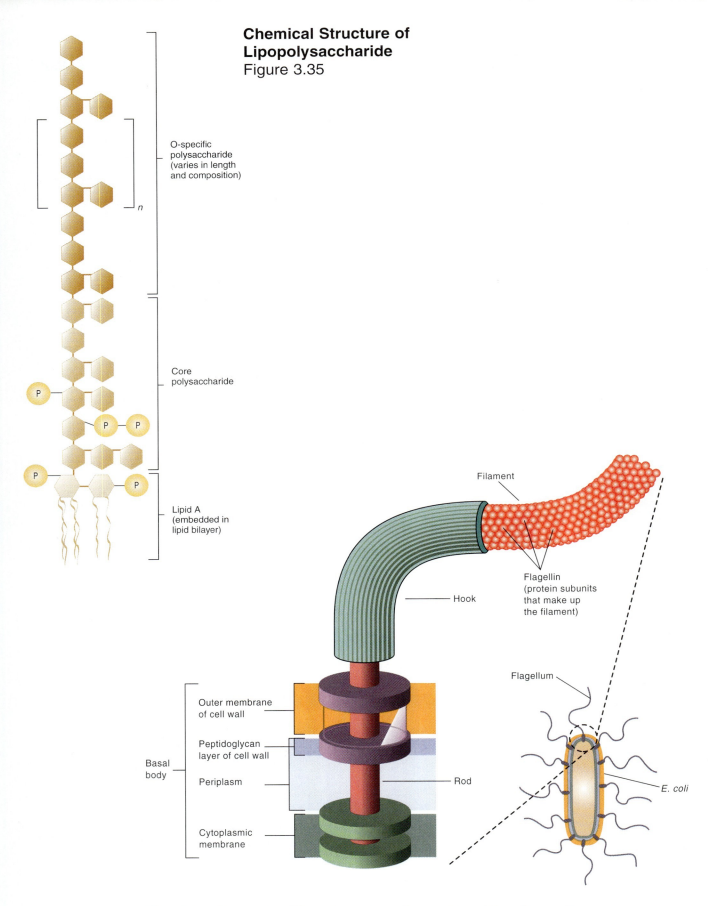

Chemical Structure of Lipopolysaccharide
Figure 3.35

The Structure of a Flagellum in a Gram-Negative Bacterium
Figure 3.39

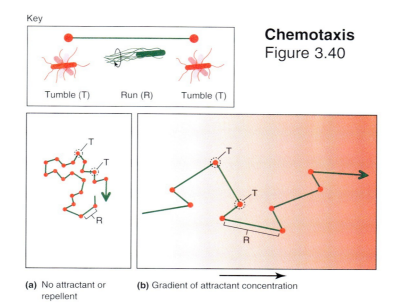

Chemotaxis
Figure 3.40

(a) No attractant or repellent
(b) Gradient of attractant concentration

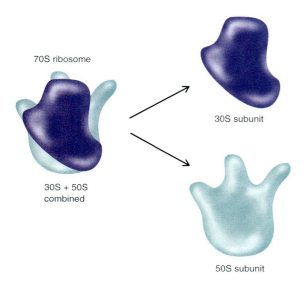

The Ribosome
Figure 3.44

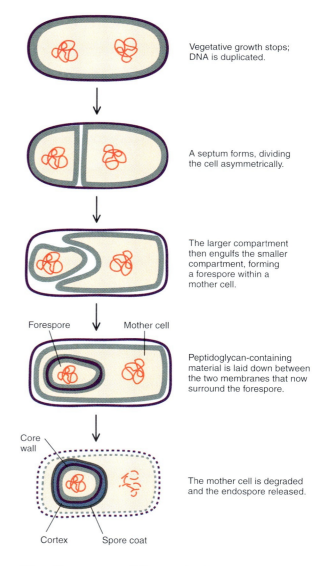

The Process of Sporulation
Figure 3.47

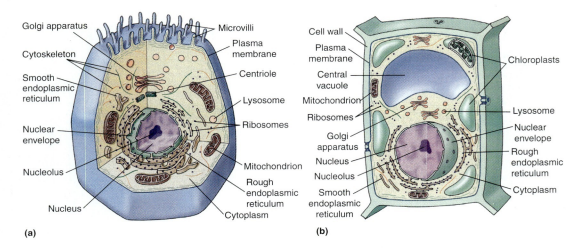

Eukaryotic Cells
Figure 3.48

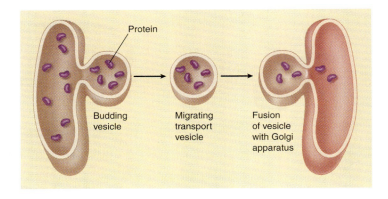

Vesicle Formation and Fusion
Figure 3.49

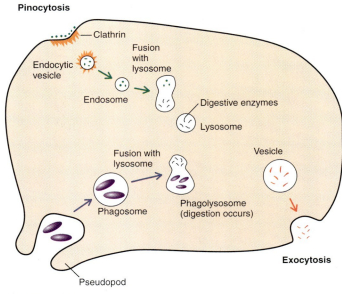

Endocytosis and Exocytosis
Figure 3.50

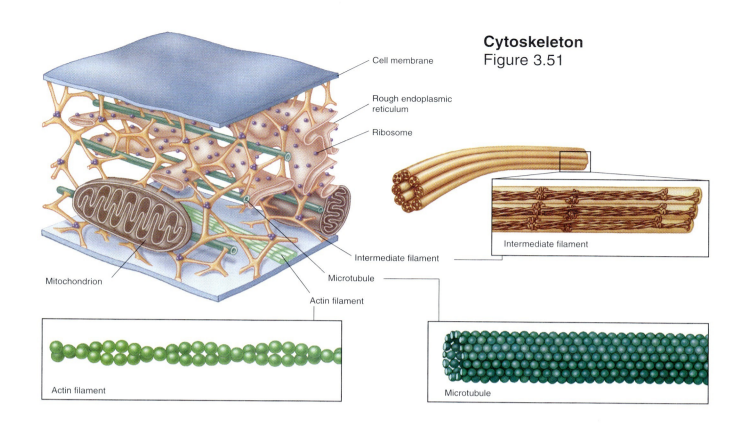

Cytoskeleton
Figure 3.51

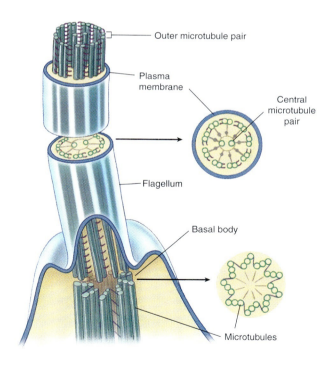

Flagella
Figure 3.52

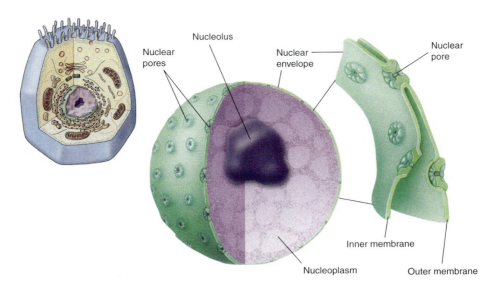

Nucleus
Figure 3.53

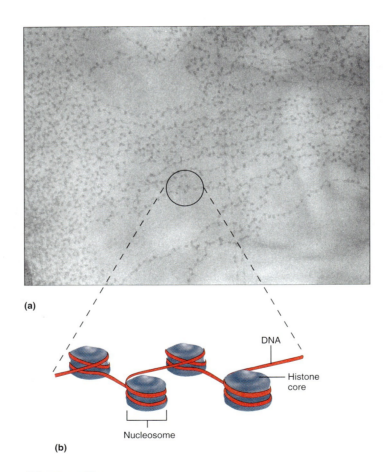

Chromatin
Figure 3.54
a: Courtesy of A. L. Olins: A. L. Olins and D. E. Olins, *Science* 183:330, 1974

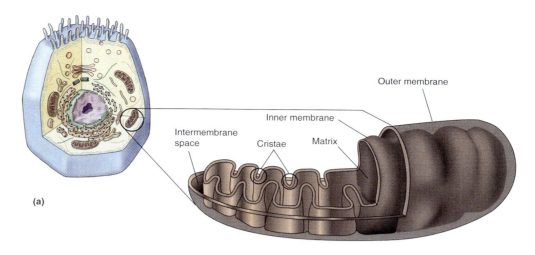

Mitochondria
Figure 3.55

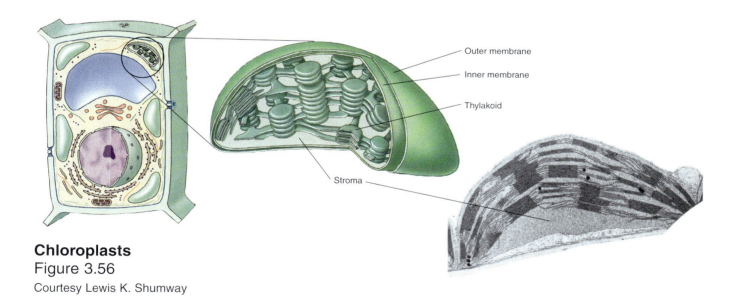

Chloroplasts
Figure 3.56
Courtesy Lewis K. Shumway

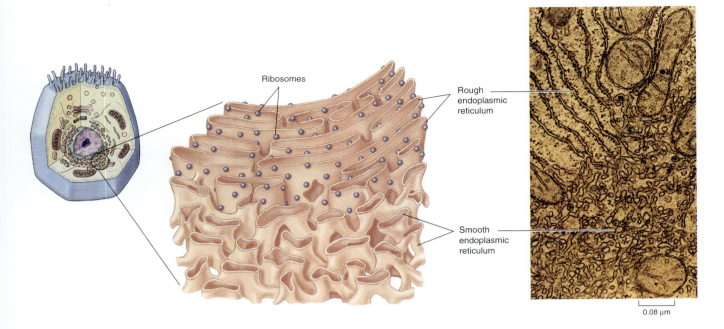

Endoplasmic Reticulum
Figure 3.57
© R. Bolendar & D. Fawcett/Visuals Unlimited

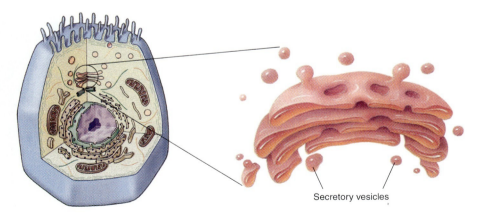

Golgi Apparatus
Figure 3.58

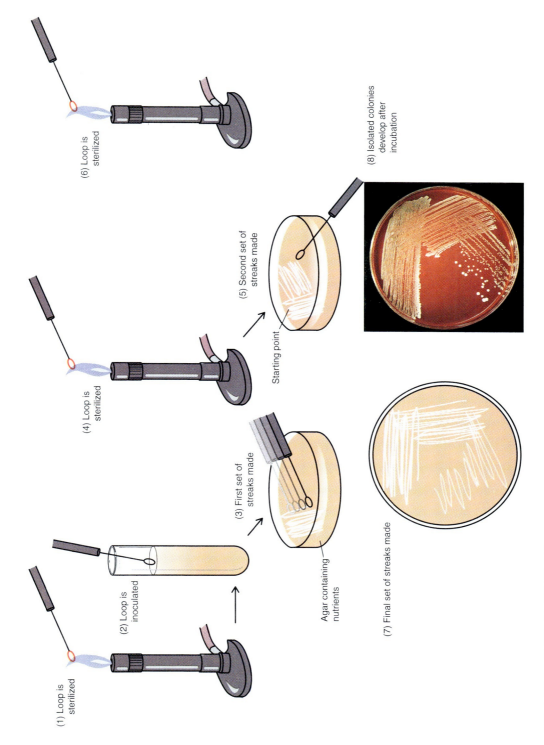

The Streak-Plate Method
Figure 4.2
© Fred E. Hossler/Visuals Unlimited

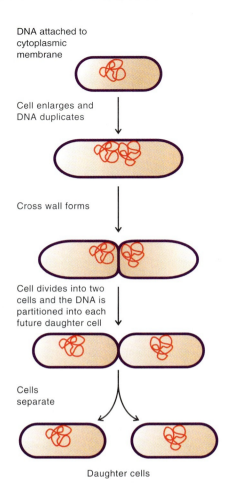

Binary Fission
Figure 4.3

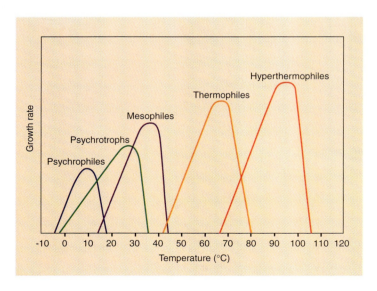

Temperature Requirements for Growth
Figure 4.4

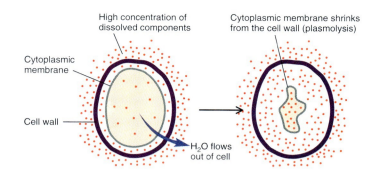

Effects of Solute Concentration on Cells
Figure 4.5

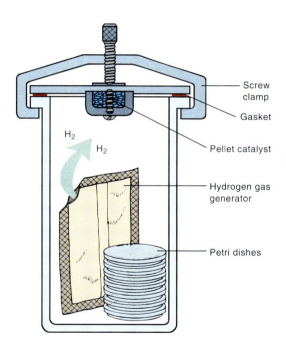

Anaerobe Jar
Figure 4.8

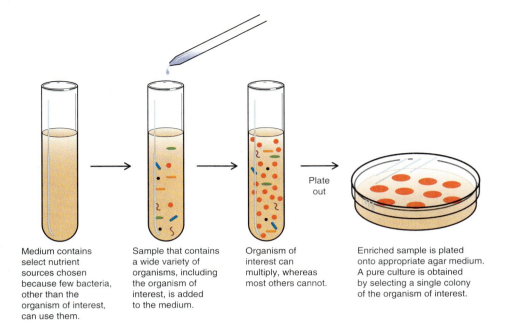

Enrichment Culture
Figure 4.10

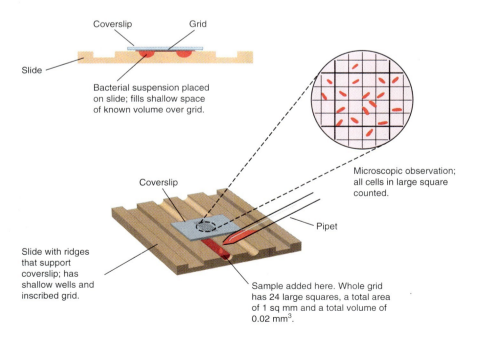

A Counting Chamber
Figure 4.11

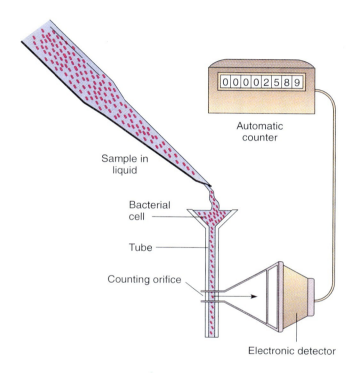

A Coulter Counter
Figure 4.12

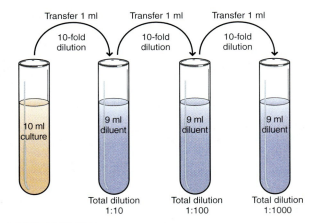
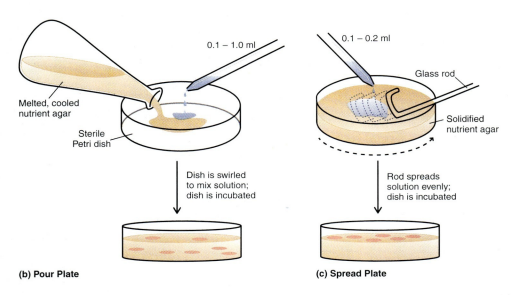

Plate Counts
Figure 4.13

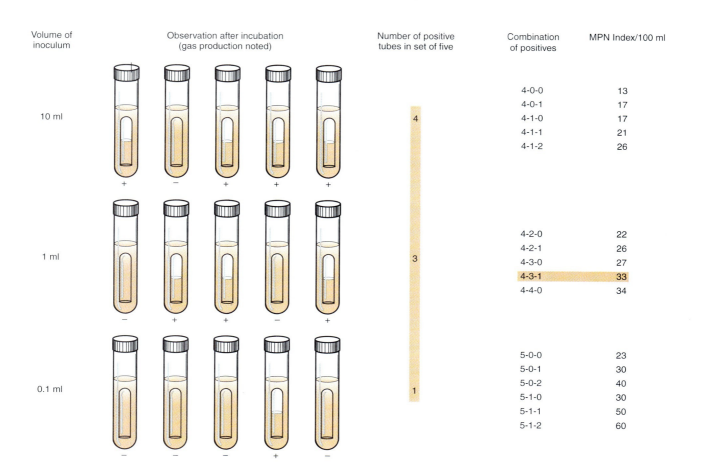

The Most Probable Number (MPN) Method
Figure 4.15

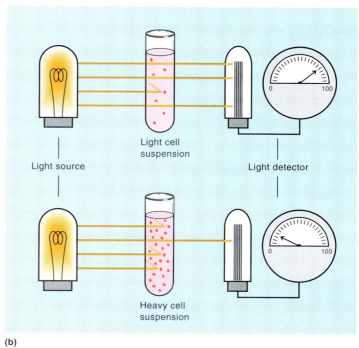

Measuring Turbidity with a Spectrophotometer
Figure 4.16

(b)

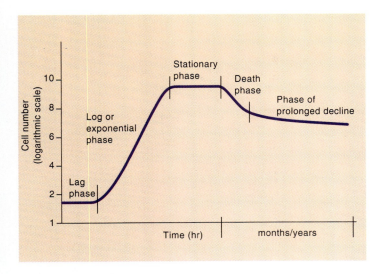

Growth Curve
Figure 4.17

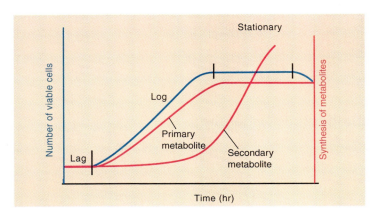

Primary and Secondary Metabolite Production
Figure 4.18

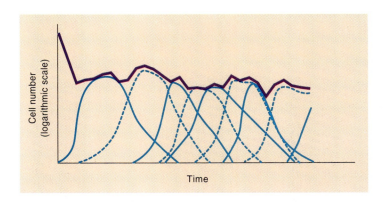

Dynamic Population Changes in the Phase of Prolonged Decline
Figure 4.19

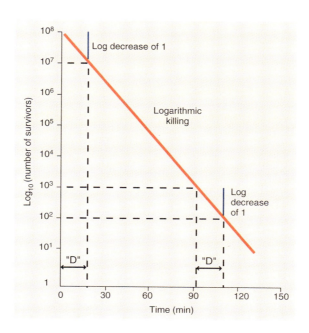

The Relationship Between the Numbers of Initial Microorganisms and the Time It Takes to Kill Them
Figure 5.2

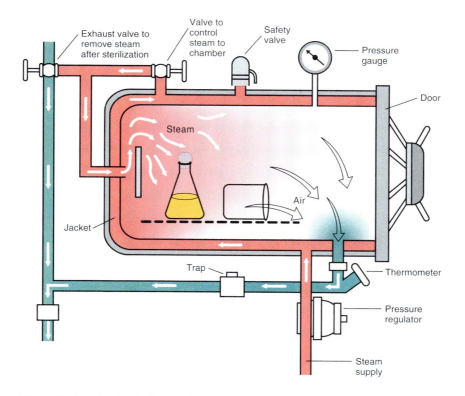

Steam-Jacketed Autoclave
Figure 5.3

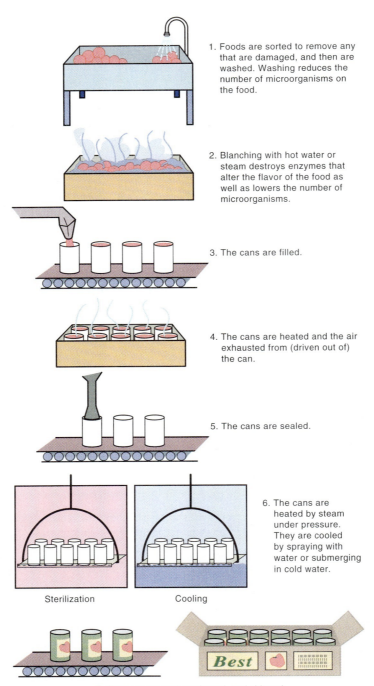

Steps in the Commercial Canning of Foods
Figure 5.5

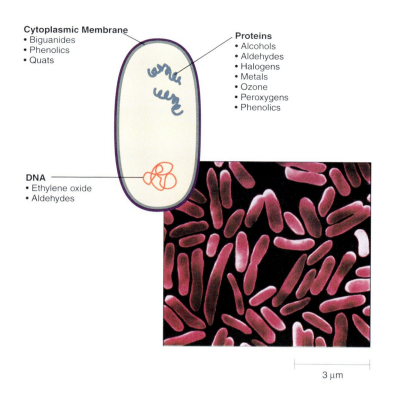

Sites of Action of Germicidal Chemical
Figure 5.6
© Dennis Kunkel/Phototake Inc.

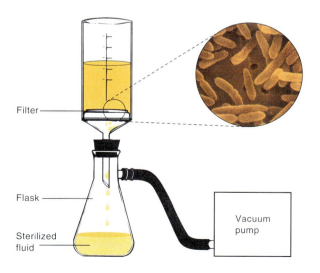

Filtration
Figure 5.7
© Pall/Visuals Unlimited

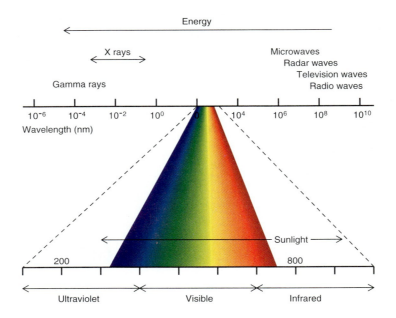

The Electromagnetic Spectrum
Figure 5.8

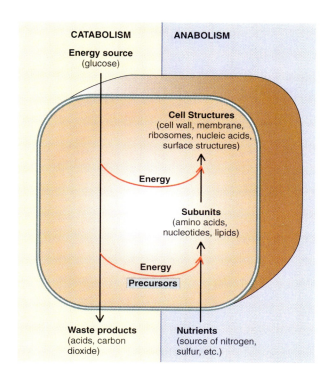

The Relationship Between Catabolism and Anabolism
Figure 6.1

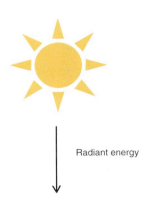

Radiant energy

Photosynthetic organisms
(harvest energy of sunlight
and use it to synthesize
organic compounds from CO_2)

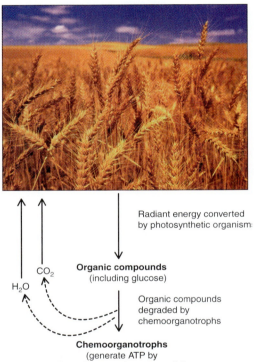

Radiant energy converted
by photosynthetic organisms

Organic compounds
(including glucose)

H_2O CO_2

Organic compounds
degraded by
chemoorganotrophs

Chemoorganotrophs
(generate ATP by
degrading organic compounds))

Most Chemoorganotrophs Depend on the Radiant Energy Harvested by Photosynthetic Organisms
Figure 6.3
top: Photodisc Vol Series 74, photo by Robert Glusie; lower: © Jane Burton/Bruce Coleman Inc.

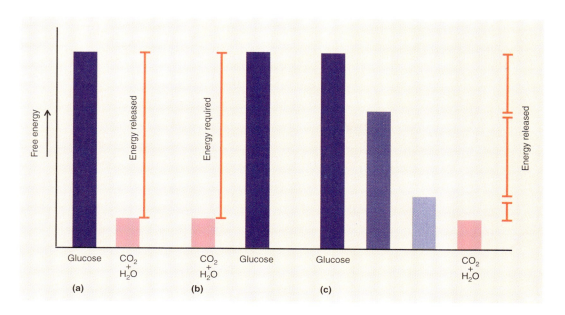

Energetics of Chemical Reactions
Figure 6.4

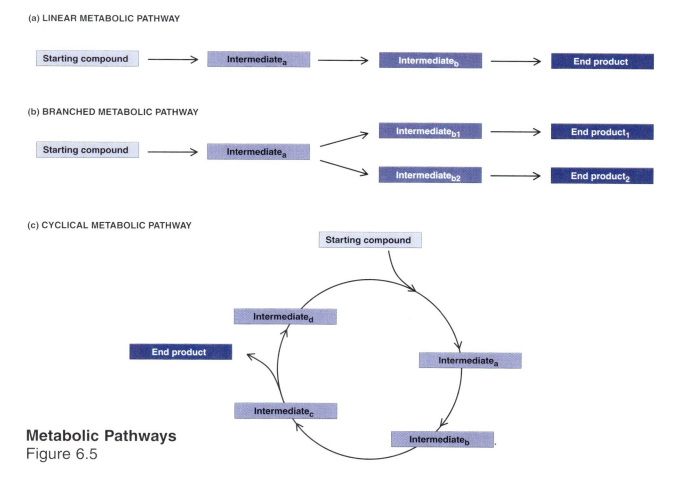

Metabolic Pathways
Figure 6.5

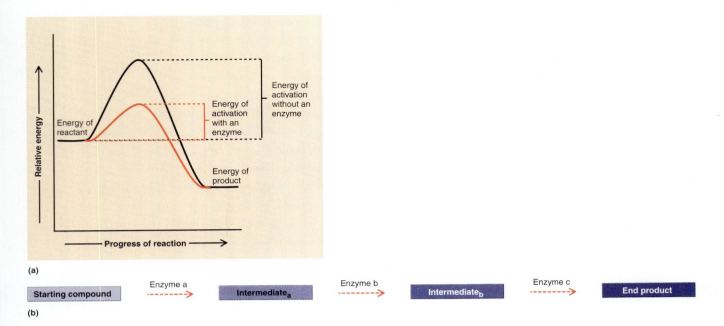

The Role of Enzymes
Figure 6.6

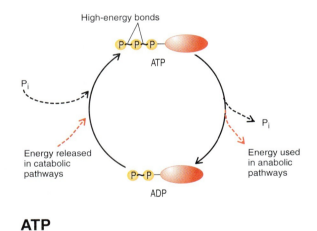

ATP
Figure 6.7

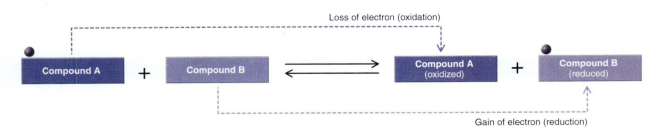

Oxidation-Reduction Reactions
Figure 6.8

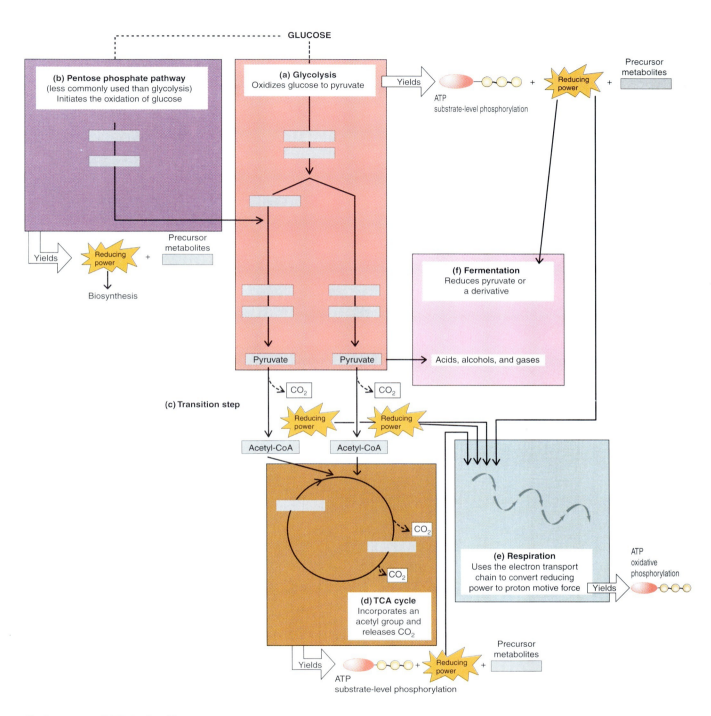

Scheme of Metabolism
Figure 6.9

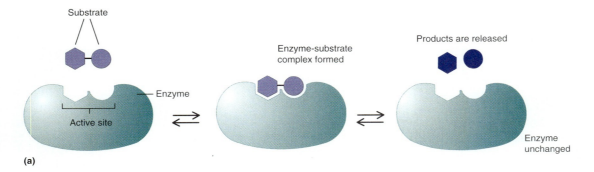

Mechanism of Enzyme Action
Figure 6.10

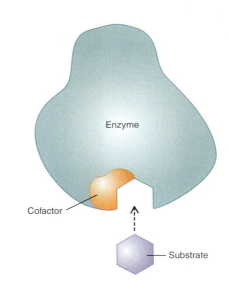

Enzymes May Act in Conjunction with a Cofactor
Figure 6.11

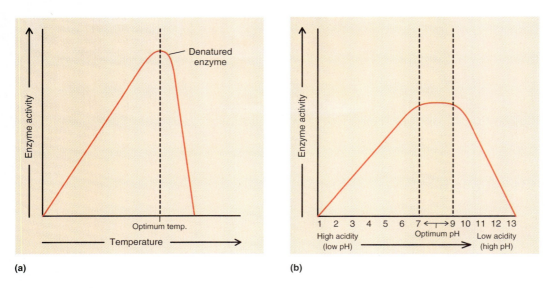

Environmental Factors that Influence Enzyme Activity
Figure 6.12

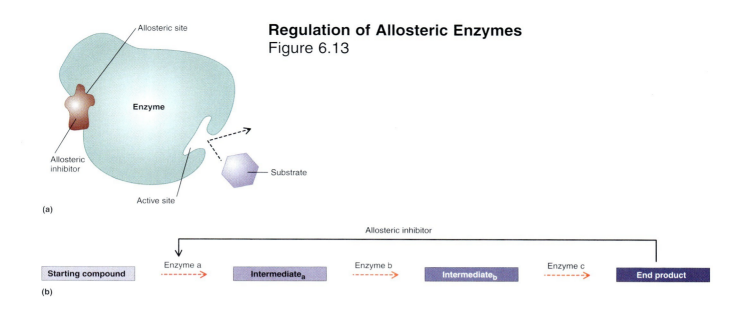

Regulation of Allosteric Enzymes
Figure 6.13

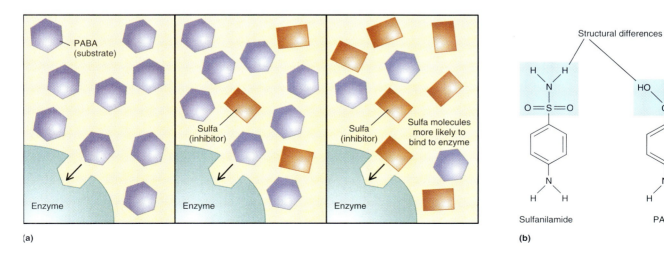

Competitive Inhibition of Enzymes
Figure 6.14

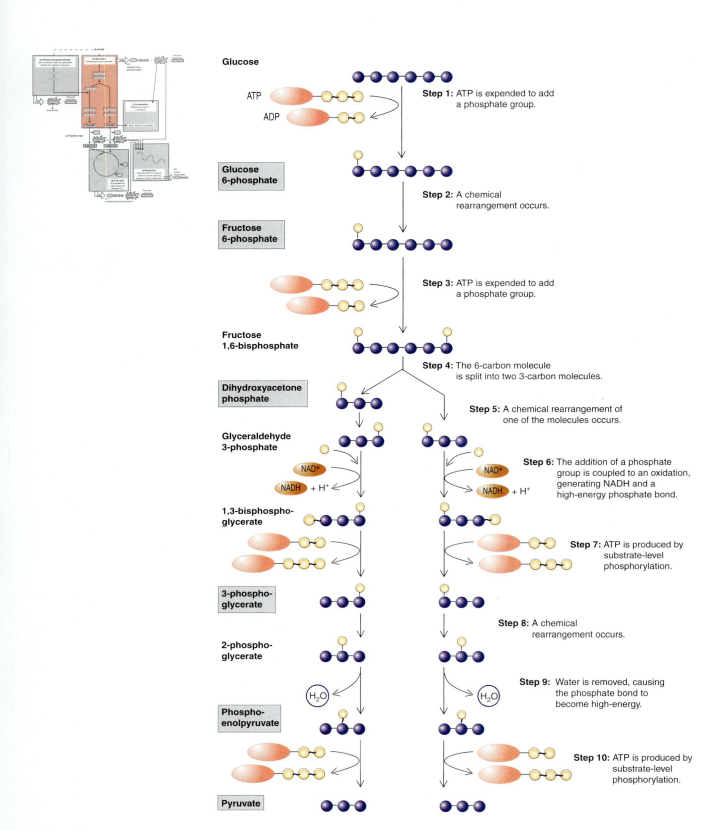

Glycolysis
Figure 6.15

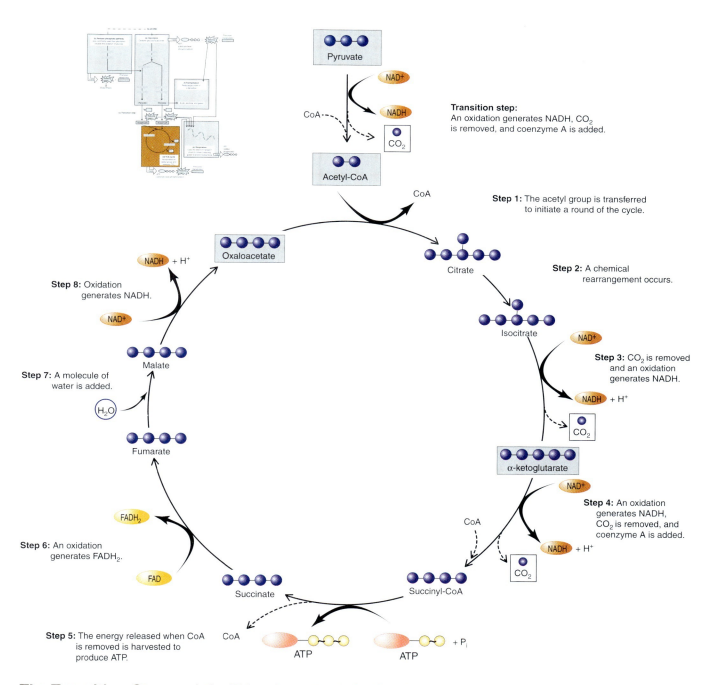

The Transition Step and the Tricarboxylic Acid Cycle
Figure 6.16

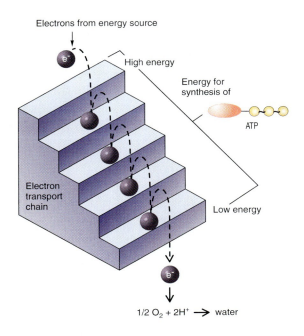

Electron Transport
Figure 6.17

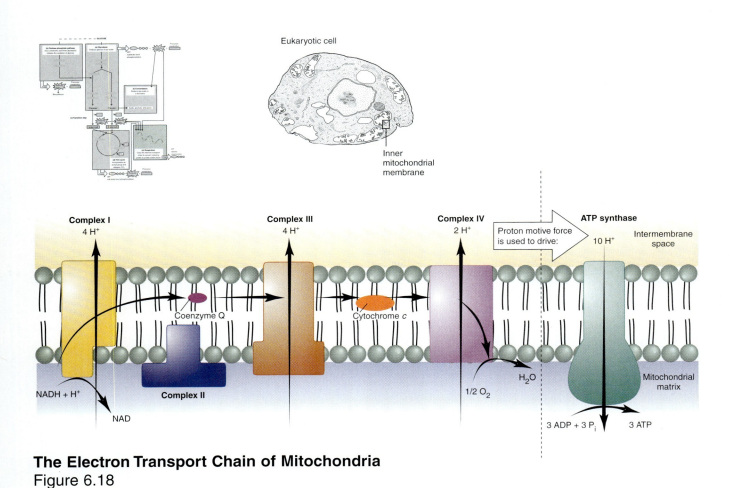

The Electron Transport Chain of Mitochondria
Figure 6.18

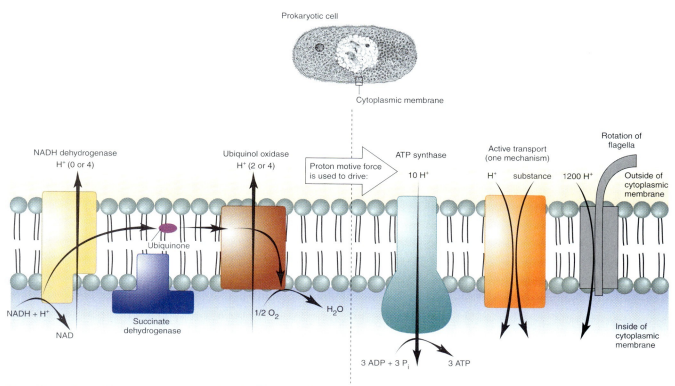

The Electron Transport Chain of *E. coli* Growing Aerobically in a Glucose-Containing Medium
Figure 6.19

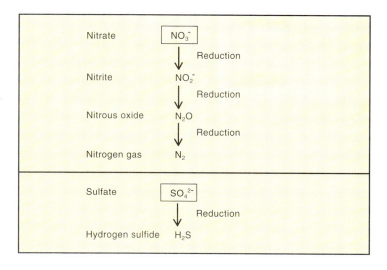

Anaerobic Respiration
Figure 6.20

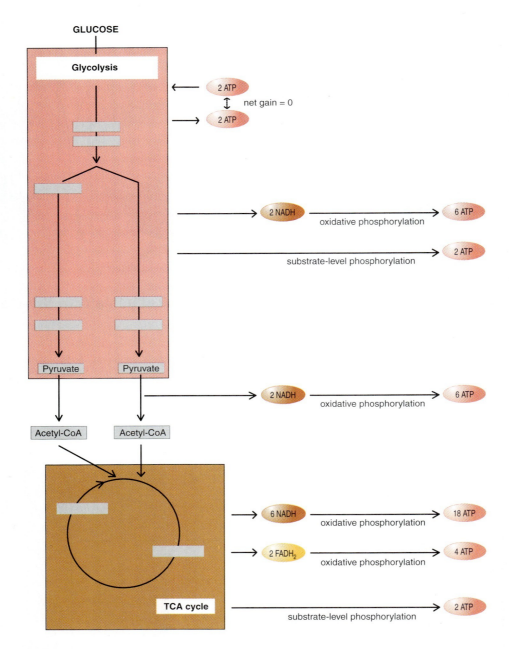

Maximum Theoretical Energy Yield from Aerobic Respiration in a Prokaryotic Cell
Figure 6.21

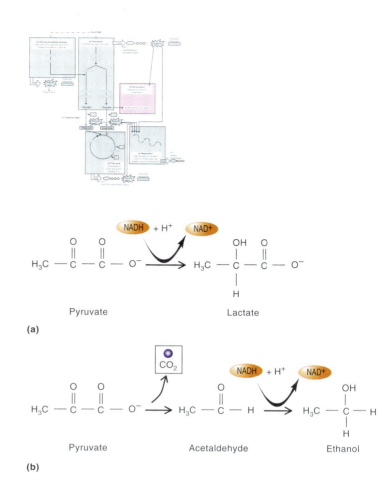

Fermentation Pathways Use Pyruvate or a Derivative As a Terminal Electron Acceptor
Figure 6.22

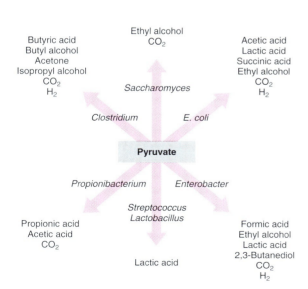

End Products of Fermentation Pathways
Figure 6.23

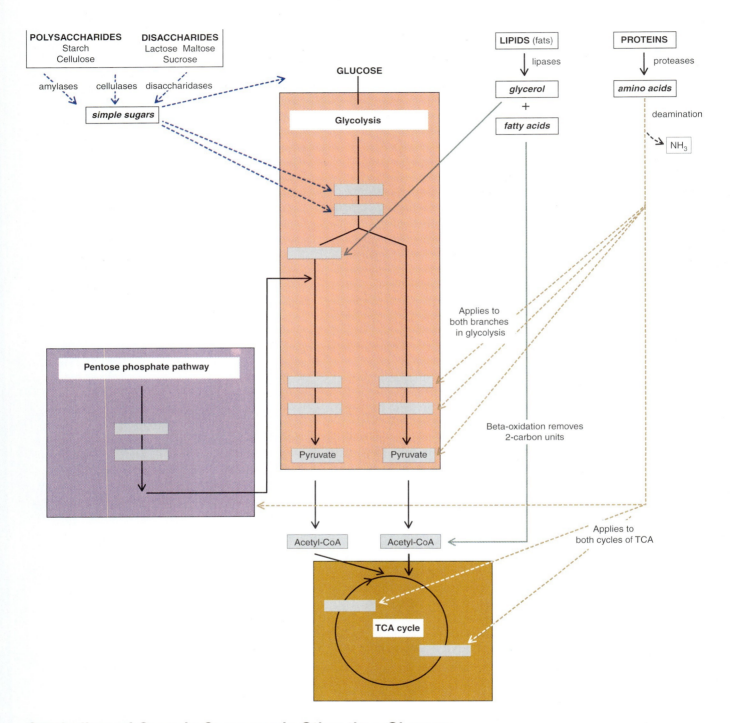

Catabolism of Organic Compounds Other than Glucose
Figure 6.24

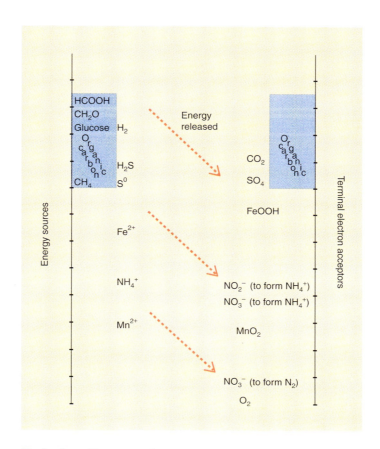

Relative Energy Gain of Different Types of Metabolism
Figure 6.25

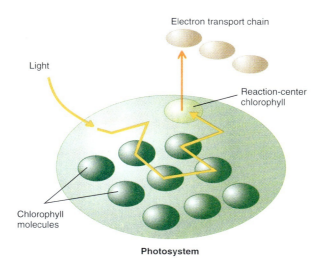

Photosystem
Figure 6.26

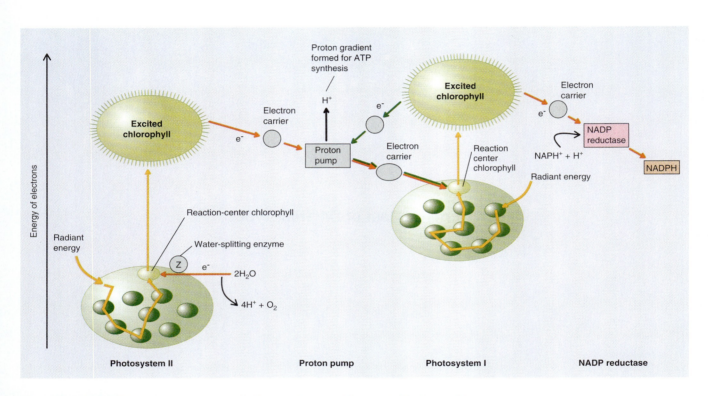

The Tandem Photosystems of Cyanobacteria and Chloroplasts
Figure 6.27

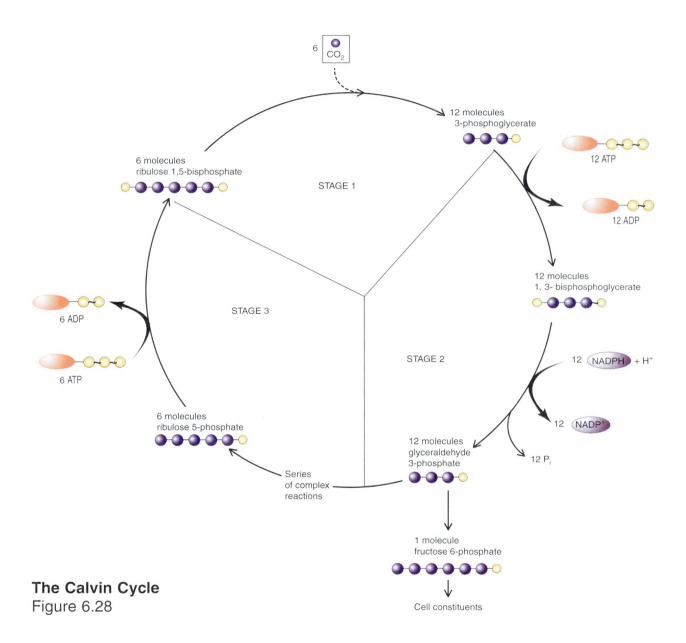

The Calvin Cycle
Figure 6.28

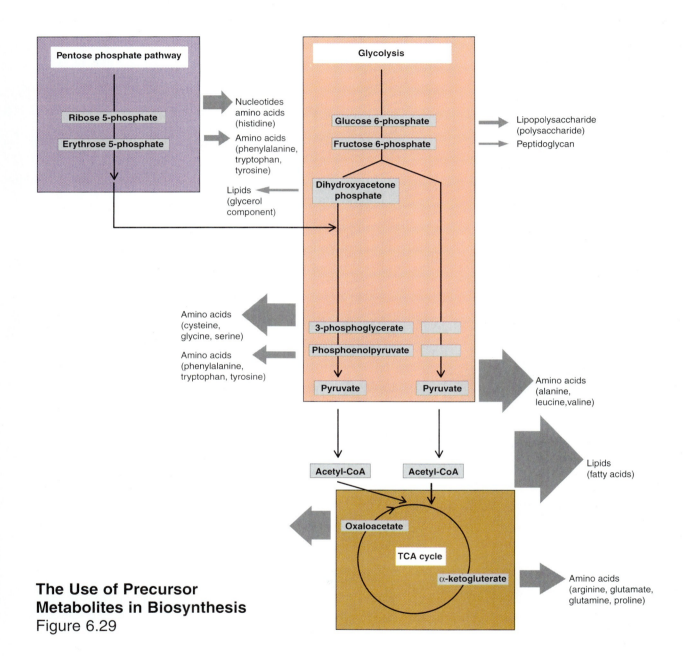

The Use of Precursor Metabolites in Biosynthesis
Figure 6.29

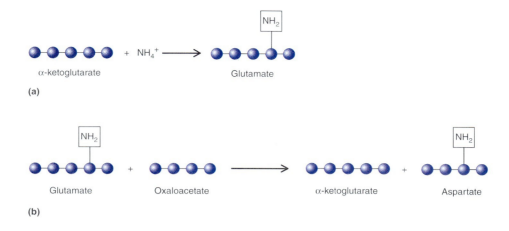

Glutamate
Figure 6.30

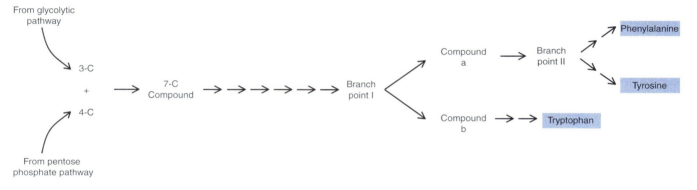

Synthesis of Aromatic Amino Acids
Figure 6.31

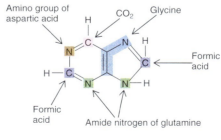

PURINE RING

Source of the Carbons and Nitrogen Atoms in Purine Rings
Figure 6.32

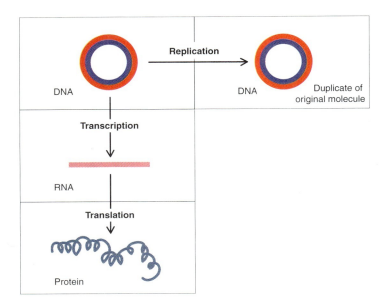

Overview of Replication, Transcription, and Translation
Figure 7.1

       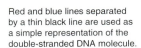

A highly coiled line is used to depict genomic DNA.

Red and blue lines placed in a helical arrangement depict the two complementary strands and highlight the three-dimensional structure of DNA.

A circular arrangement of the red and blue lines is used as the simplified form of prokaryotic DNA.

Red and blue lines separated by a thin black line are used as a simple representation of the double-stranded DNA molecule.

Base-pairing

Two parallel lines are used to emphasize the base-pairing interactions and nucleotide sequence characteristics of the two complementary strands. The "tracks" between the lines are not intended to depict a specific number of base pairs, only the general interaction between complementary strands.

Either a red or a blue line can be depicted as the "top" strand, since DNA is a three-dimensional structure.

Denatured DNA is depicted as separate red and blue lines to emphasize its single-stranded nature.

Diagrammatic Representations of the Structure of DNA
Figure 7.2

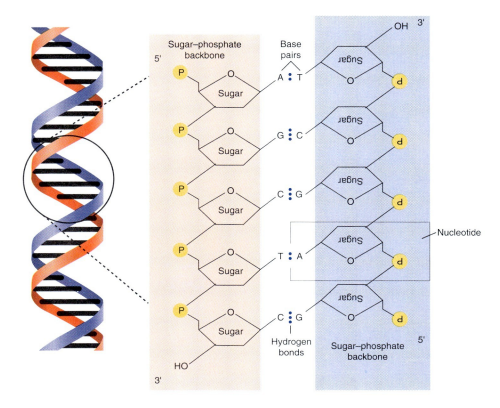

The Double Helix of DNA
Figure 7.3

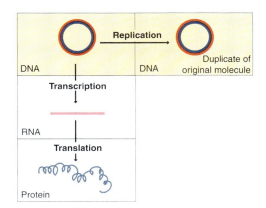

Replication of Chromosomal DNA of Prokaryotes
Figure 7.4

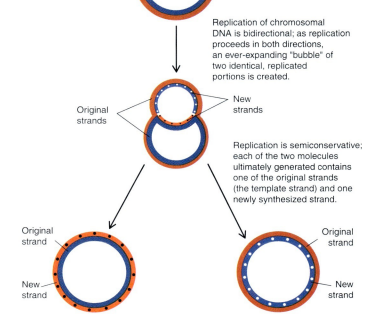

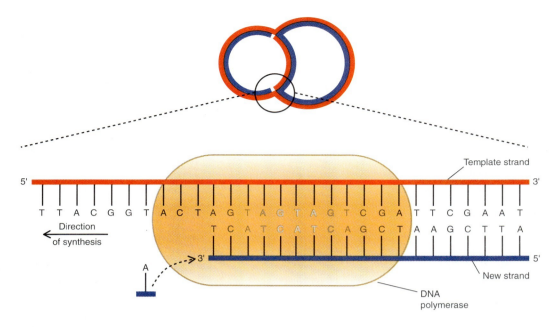

The Process of DNA Synthesis
Figure 7.5

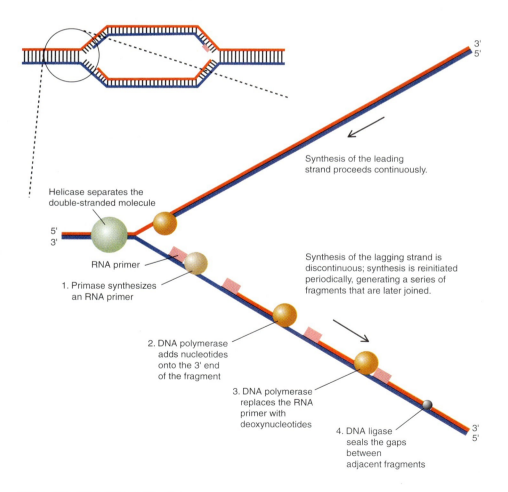

The Replication Fork
Figure 7.6

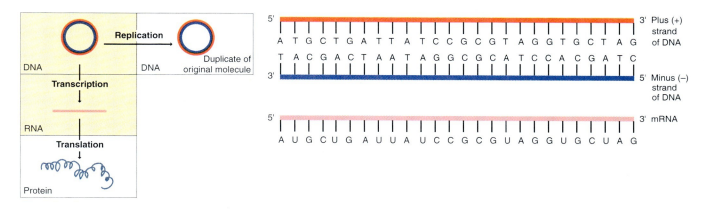

RNA Is Transcribed from a DNA Template
Figure 7.7

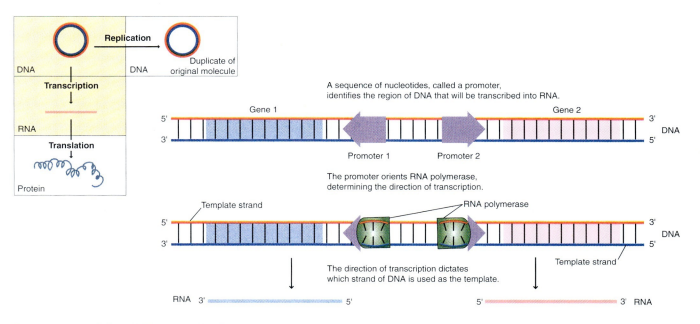

Promoters Direct Transcription
Figure 7.8

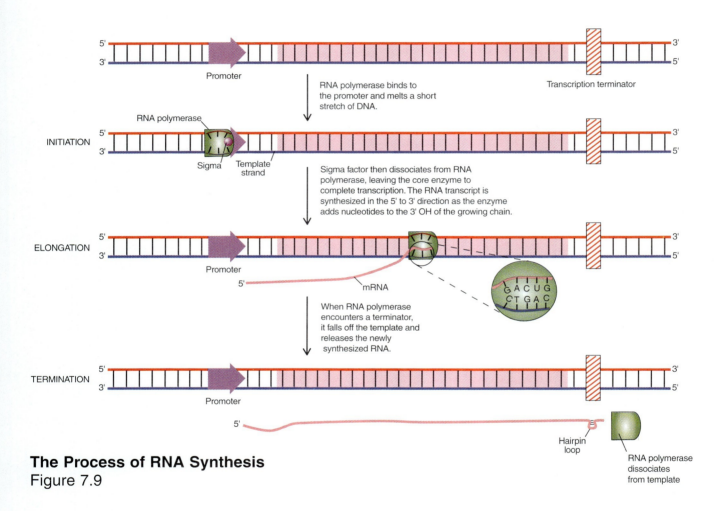

The Process of RNA Synthesis
Figure 7.9

The Genetic Code
Figure 7.10

First Letter	Middle Letter U Reading frame 5' 3'		Middle Letter C Reading frame 5' 3'		Middle Letter A Reading frame 5' 3'		Middle Letter G Reading frame 5' 3'		Last Letter
U	UUU	Phenylalanine	UCU	Serine	UAU	Tyrosine	UGU	Cysteine	U
	UUC	Phenylalanine	UCC	Serine	UAC	Tyrosine	UGC	Cysteine	C
	UUA	Leucine	UCA	Serine	UAA	(Stop)	UGA	(Stop)	A
	UUG	Leucine	UCG	Serine	UAG	(Stop)	UGG	Tryptophan	G
C	CUU	Leucine	CCU	Proline	CAU	Histidine	CGU	Arginine	U
	CUC	Leucine	CCC	Proline	CAC	Histidine	CGC	Arginine	C
	CUA	Leucine	CCA	Proline	CAA	Glutamine	CGA	Arginine	A
	CUG	Leucine	CCG	Proline	CAG	Glutamine	CGG	Arginine	G
A	AUU	Isoleucine	ACU	Threonine	AAU	Asparagine	AGU	Serine	U
	AUC	Isoleucine	ACC	Threonine	AAC	Asparagine	AGC	Serine	C
	AUA	Isoleucine	ACA	Threonine	AAA	Lysine	AGA	Arginine	A
	AUG	Methionine (Start)	ACG	Threonine	AAG	Lysine	AGG	Arginine	G
G	GUU	Valine	GCU	Alanine	GAU	Aspartate	GGU	Glycine	U
	GUC	Valine	GCC	Alanine	GAC	Aspartate	GGC	Glycine	C
	GUA	Valine	GCA	Alanine	GAA	Glutamate	GGA	Glycine	A
	GUG	Valine	GCG	Alanine	GAG	Glutamate	GGG	Glycine	G

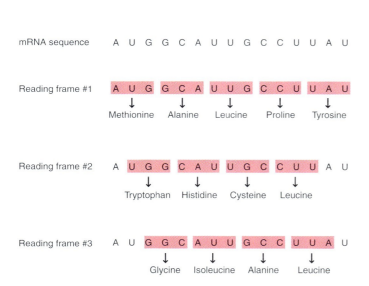

Reading Frames
Figure 7.11

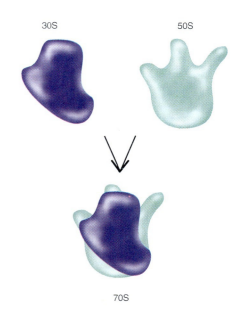

The Structure of the 70S Ribosome
Figure 7.12

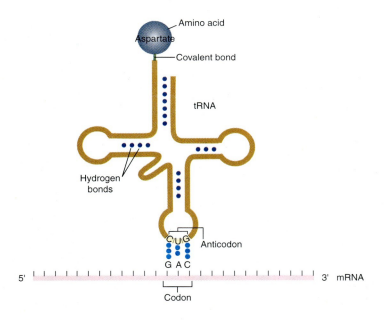

The Structure of Transfer RNA (tRNA)
Figure 7.13

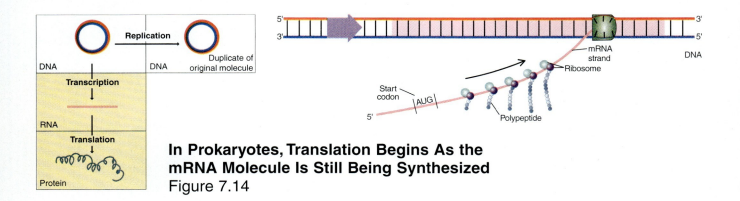

In Prokaryotes, Translation Begins As the mRNA Molecule Is Still Being Synthesized
Figure 7.14

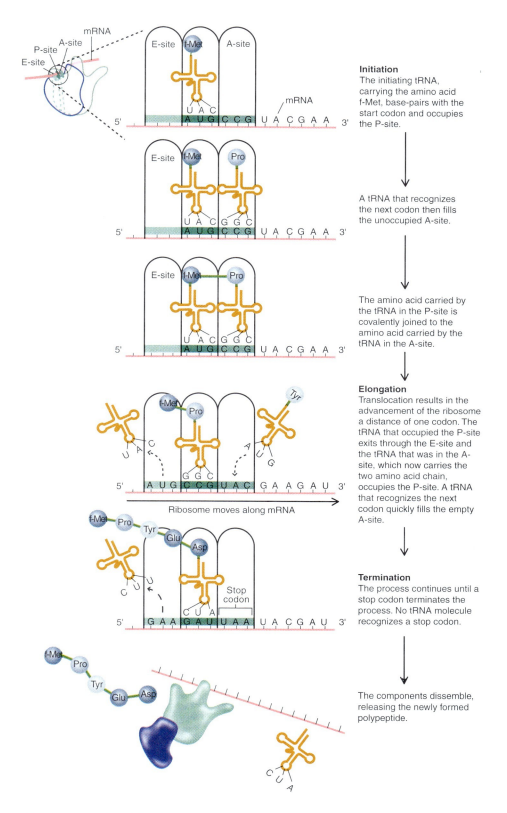

The Process of Translation
Figure 7.15

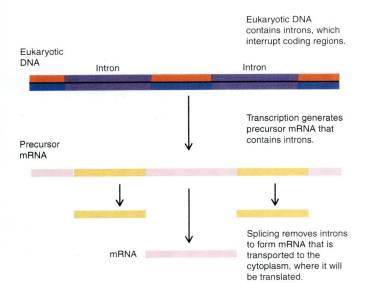

Splicing of Eukaryotic RNA
Figure 7.16

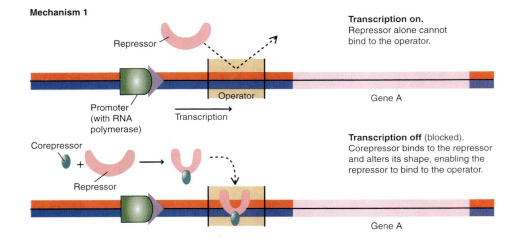

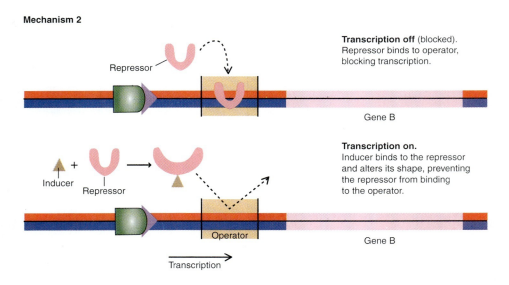

Transcriptional Regulation by Repressors
Figure 7.18

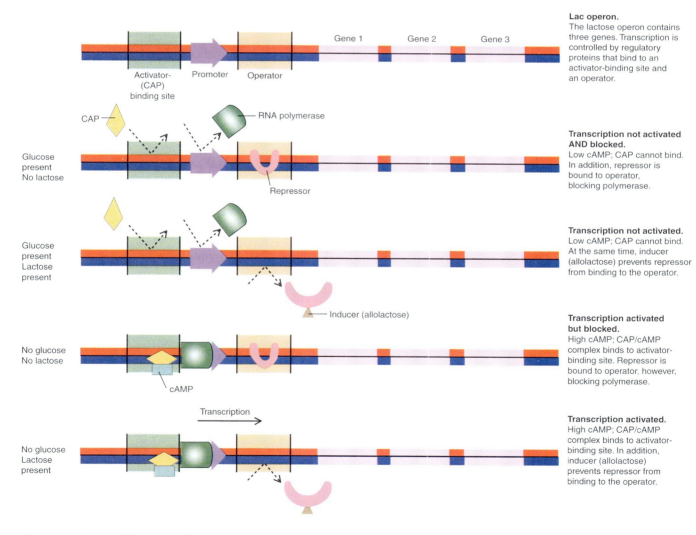

Transcriptional Regulation by Activators
Figure 7.19

Regulation of the *lac* Operon
Figure 7.20

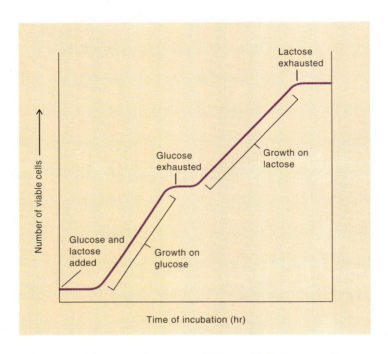

Diauxic Growth Curve of *E. coli* Growing in a Medium Containing Glucose and Lactose
Figure 7.21

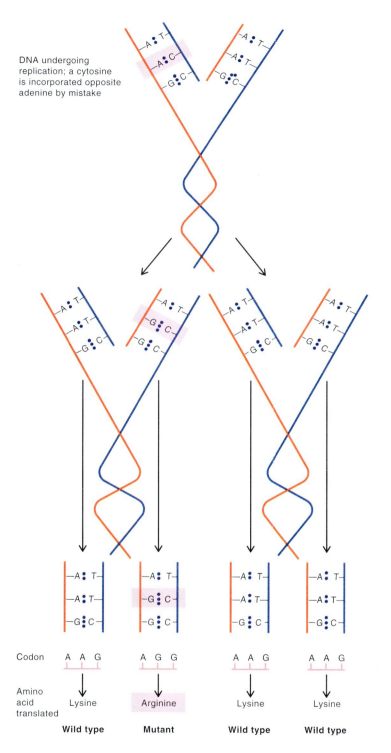

Base Substitution
Figure 8.1

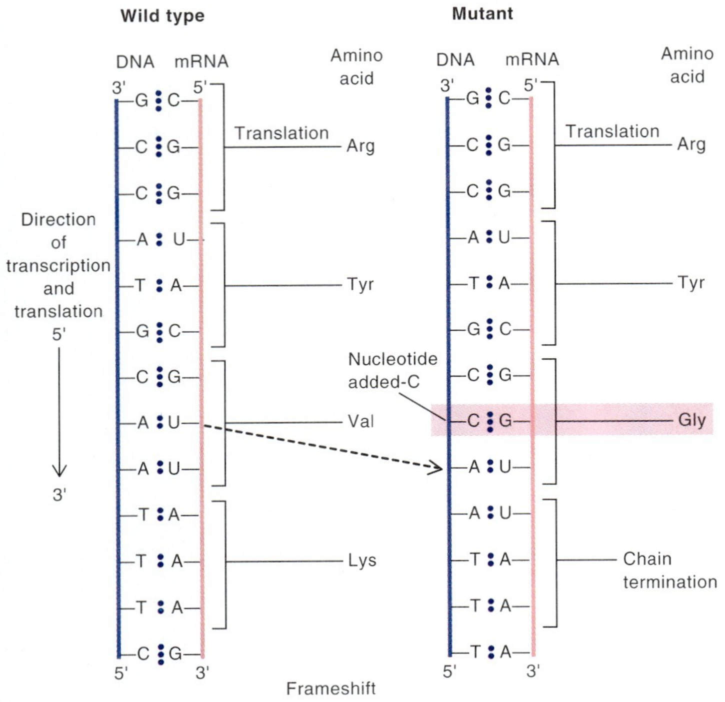

Frameshift Mutation As a Result of Base Addition
Figure 8.2

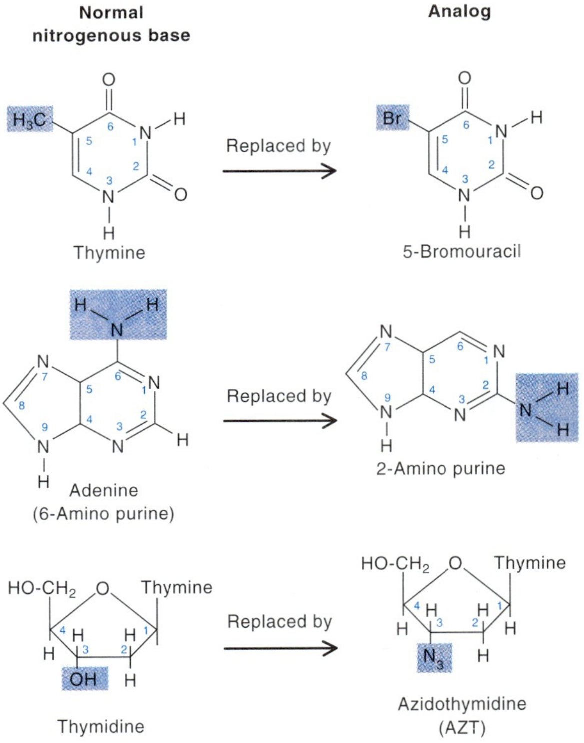

Common Base Analogs and the Normal Bases They Replace in DNA
Figure 8.4

Thymine Dimer Formation
Figure 8.5

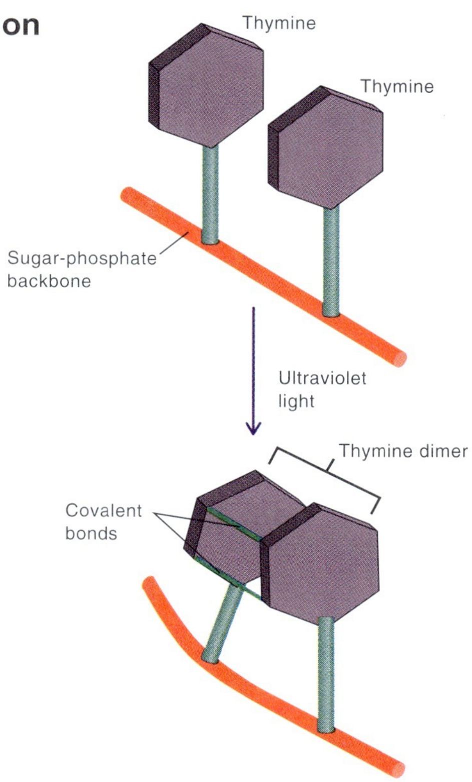

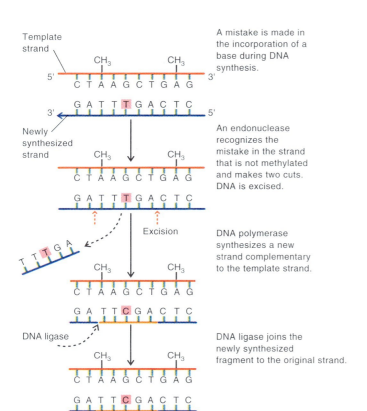

Mismatch Repair
Figure 8.6

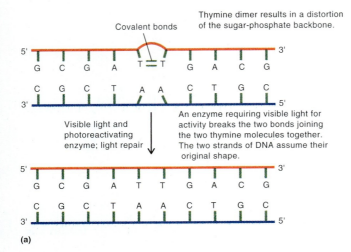
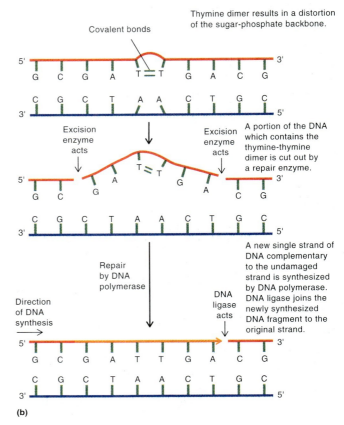

Repair of Thymine Dimers
Figure 8.7

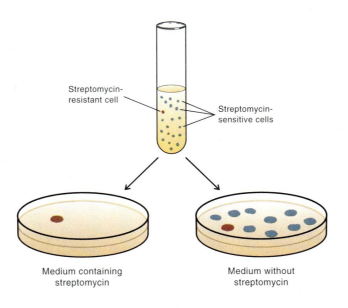

Direct Selection of Mutants
Figure 8.8

Indirect Selection of Mutants by Replica Plating
Figure 8.9

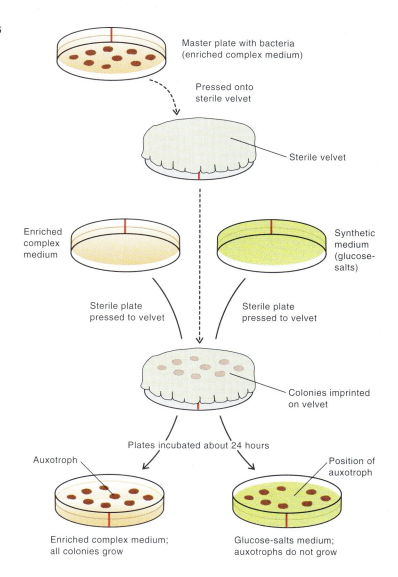

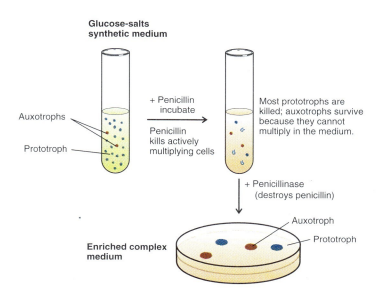

Penicillin Enrichment of Mutants
Figure 8.10

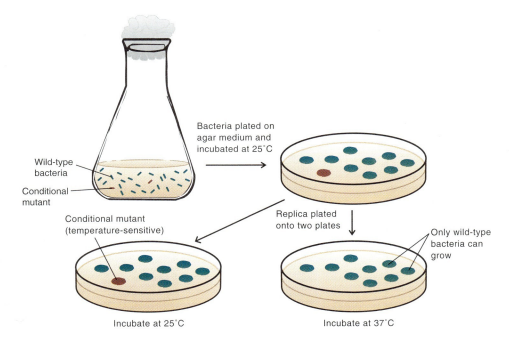

Isolation of Temperature-Sensitive (Conditional) Mutants
Figure 8.11

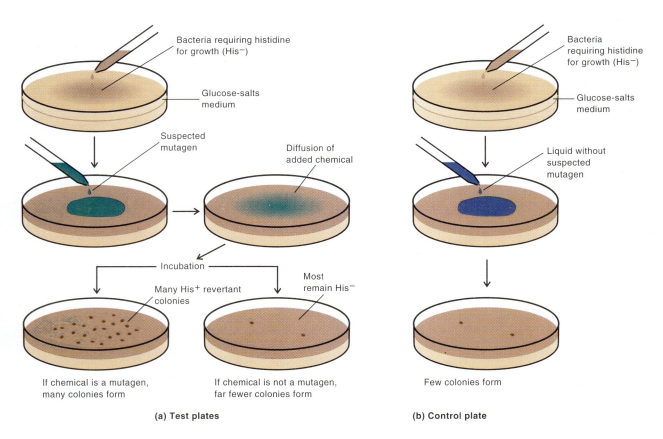

Ames Test to Screen for Mutagens
Figure 8.12

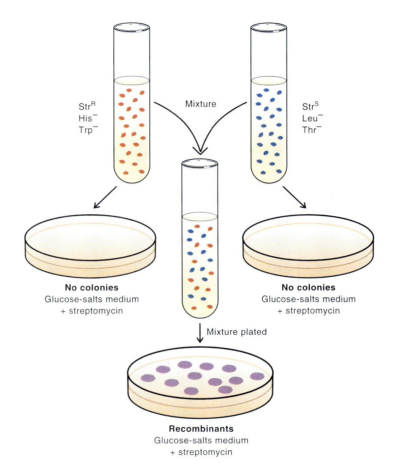

General Experimental Approach for Detecting Gene Transfer in Bacteria
Figure 8.13

DNA-Mediated Transformation
Figure 8.14

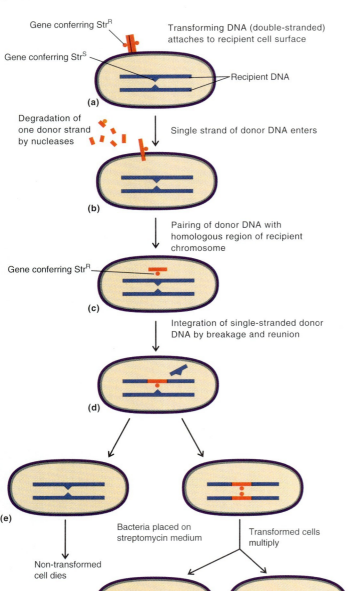

Electroporation
Figure 8.15

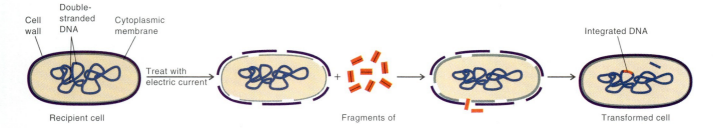

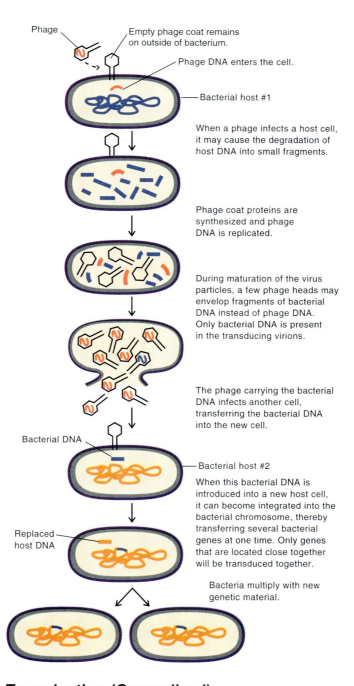

Transduction (Generalized)
Figure 8.16

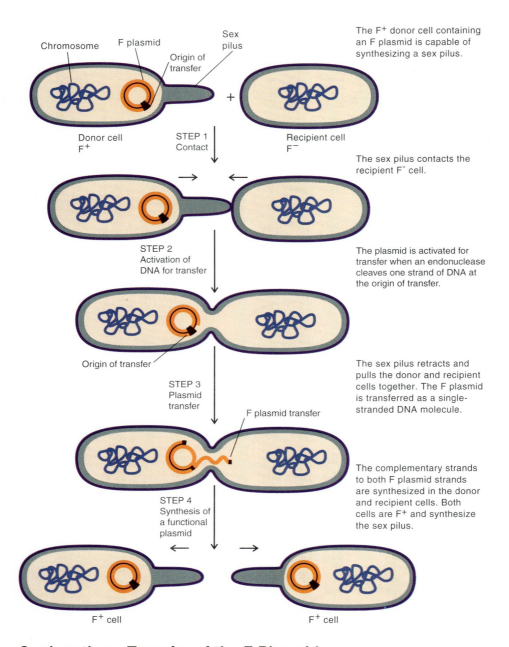

Conjugation—Transfer of the F Plasmid
Figure 8.18

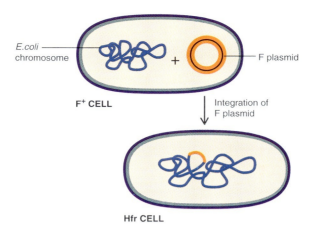

Hfr Formation
Figure 8.19

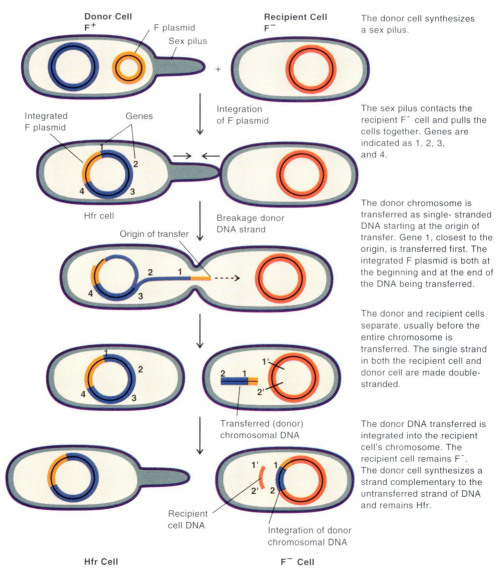

Conjugation—Transfer of Chromosomal DNA
Figure 8.20

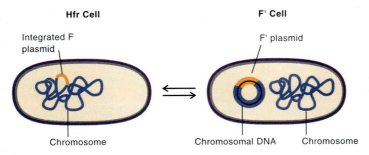

Formation of F′ Plasmid
Figure 8.21

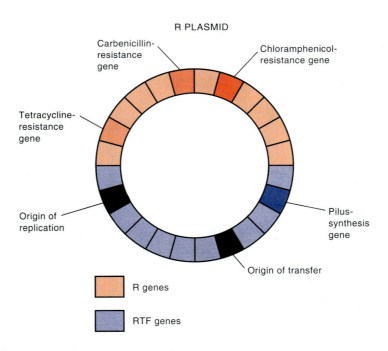

Two Regions of an R Plasmid
Figure 8.22

The Movement of a Transposon Throughout a Bacterial Population
Figure 8.23

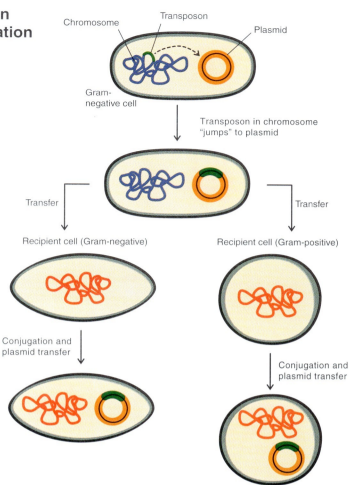

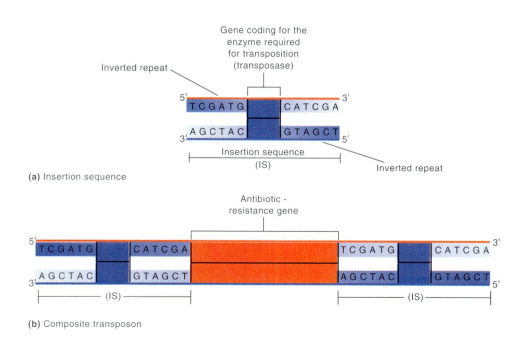

(a) Insertion sequence

(b) Composite transposon

Transposable Elements
Figure 8.24

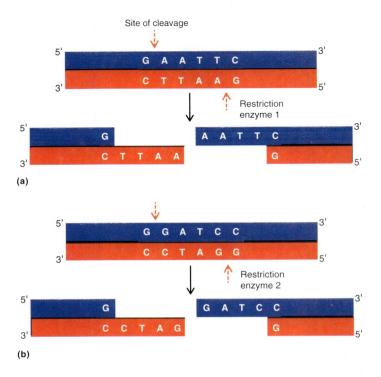

Action of Restriction Endonucleases
Figure 8.25

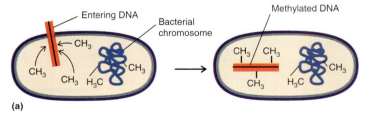

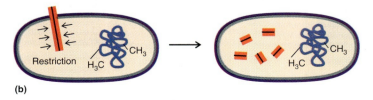

Restriction and Modification of Entering DNA
Figure 8.26

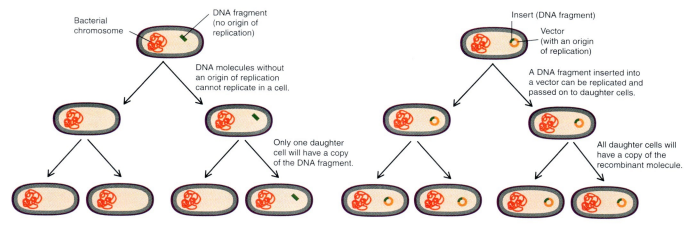

DNA Must Replicate in a Cell in Order to be Maintained in a Population of Cells
Figure 9.1

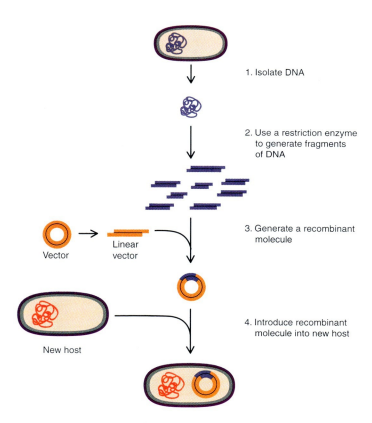

The Steps of a Cloning Experiment
Figure 9.2

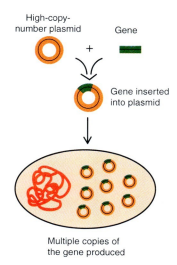

Cloning into a High-Copy-Number Plasmid
Figure 9.3

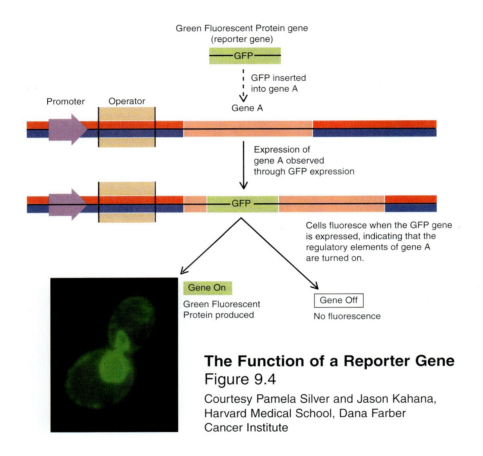

The Function of a Reporter Gene
Figure 9.4
Courtesy Pamela Silver and Jason Kahana, Harvard Medical School, Dana Farber Cancer Institute

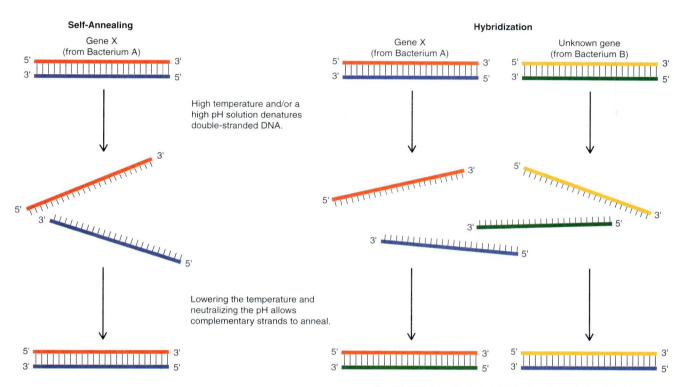

Nucleic Acid Hybridization Can Be Used to Locate Homologous Sequences
Figure 9.6

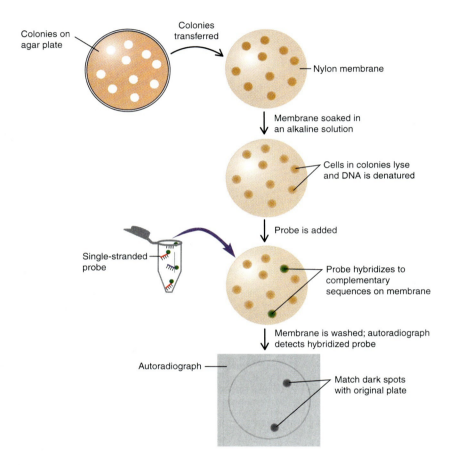

The Steps of a Colony Blot
Figure 9.7

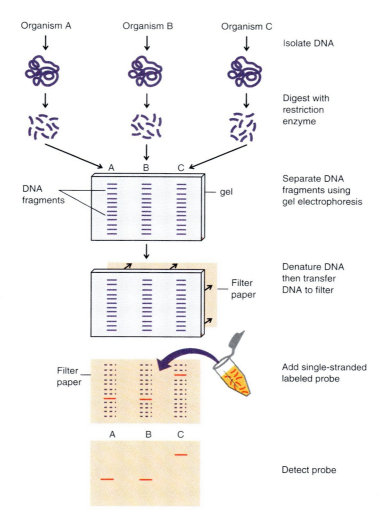

The Steps of a Southern Blot
Figure 9.8

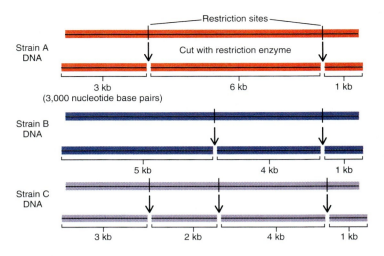

Restriction Fragment Length Polymorphism (RFLP)
Figure 9.9

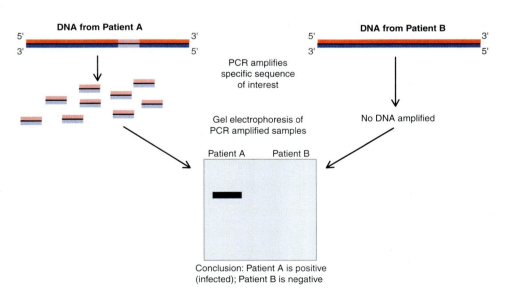

PCR Amplifies Selected Sequences
Figure 9.11

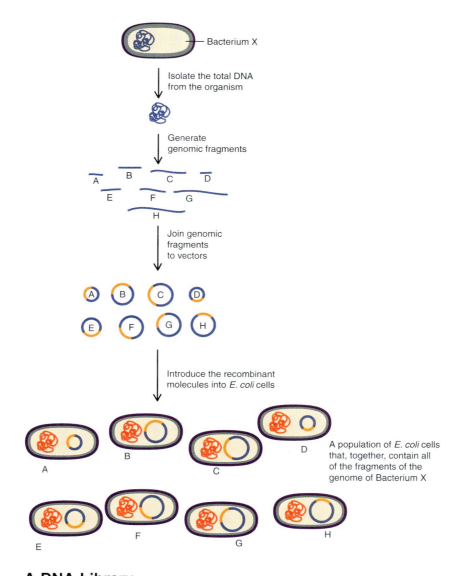

A DNA Library
Figure 9.12

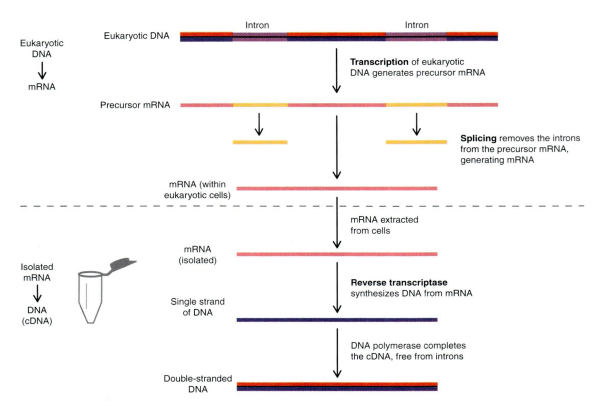

Making cDNA from Eukaryotic mRNA
Figure 9.13

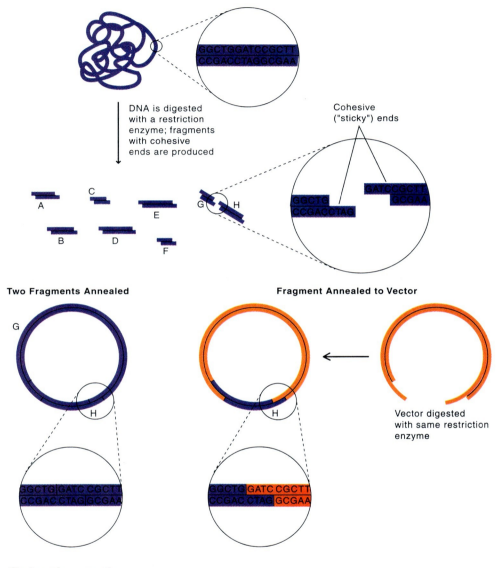

Cohesive ends
Figure 9.14

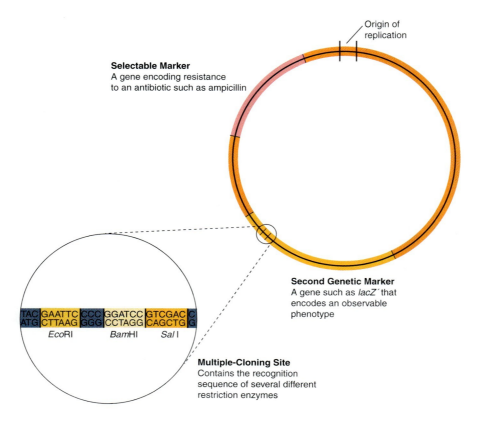

Typical Properties of an Ideal Vector
Figure 9.15

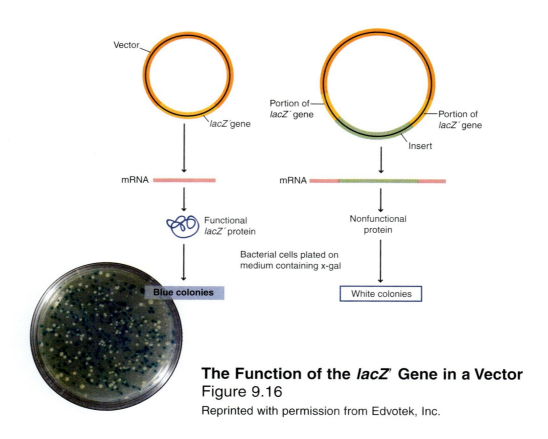

The Function of the *lacZ'* Gene in a Vector
Figure 9.16
Reprinted with permission from Edvotek, Inc.

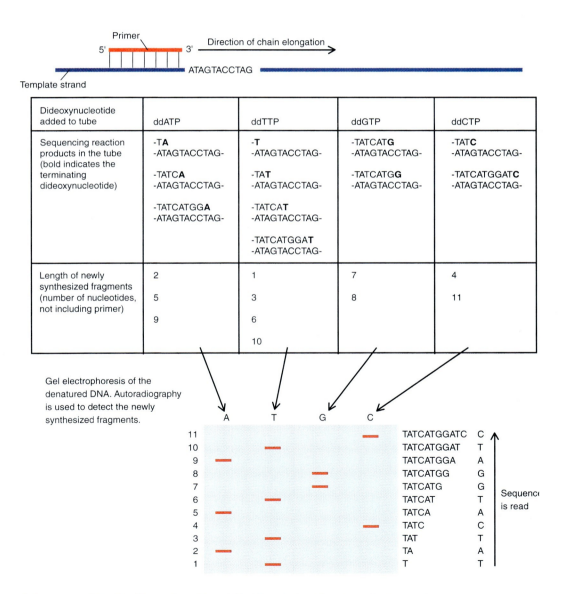

Chain Termination by a Dideoxynucleotide
Figure 9.19

Dideoxy Chain Termination Method for Determining the Nucleotide Sequence of DNA
Figure 9.20

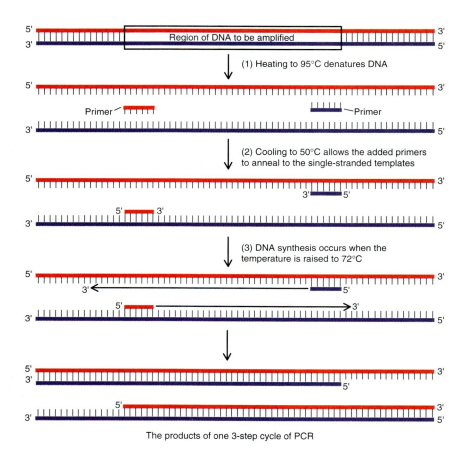

Steps of a Single Cycle of PCR
Figure 9.23

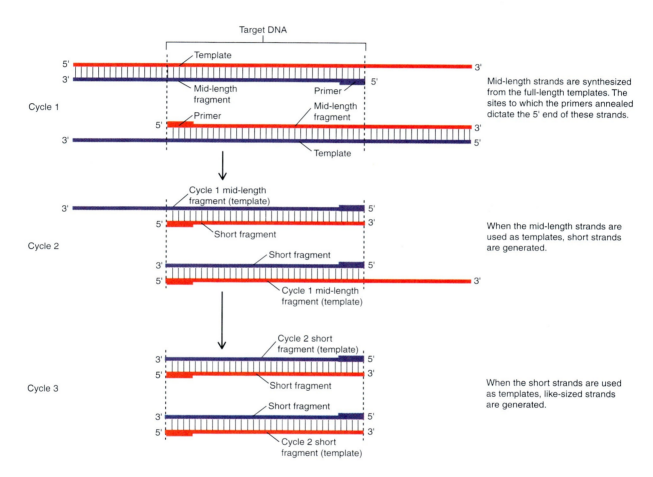

The Final PCR Product Is a Fragment of Discrete Size
Figure 9.24

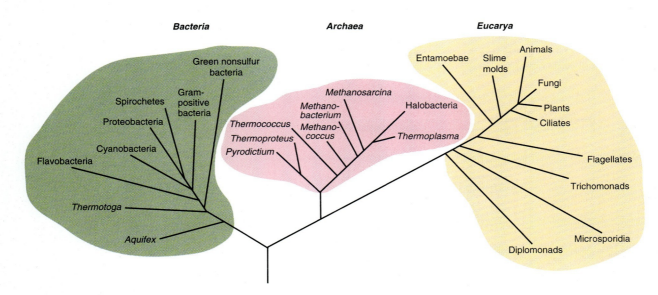

The Three-Domain System of Classification
Figure 10.1

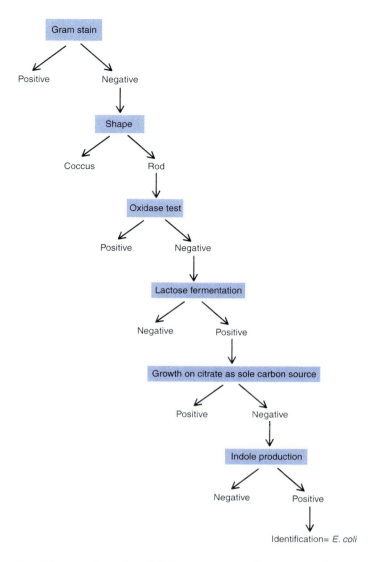

An Example of a Dichotomous Key Leading to the Identification of *E. coli*
Figure 10.5

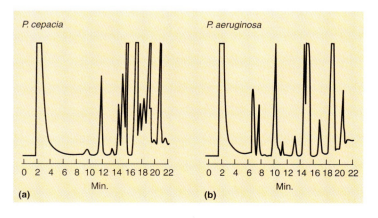

Chromatogram of the Fatty Acid Profiles
Figure 10.7

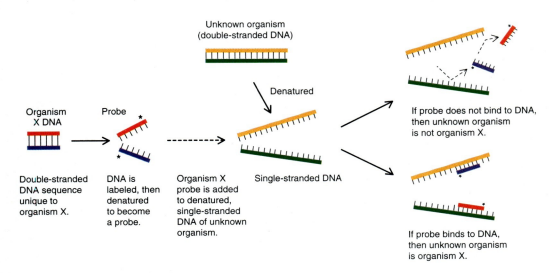

Nucleic Acid Probes to Detect Specific DNA Sequences
Figure 10.8

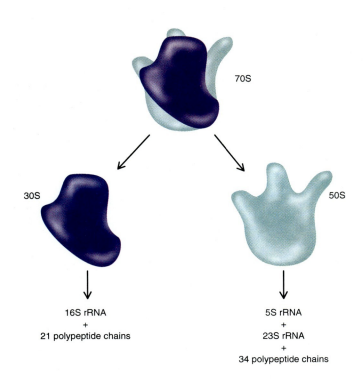

Ribosomal RNA
Figure 10.9

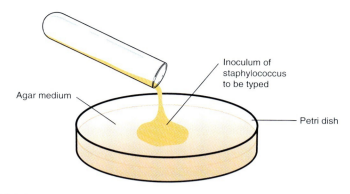

(a) An inoculum of *S. aureus* is spread over the surface of agar medium.

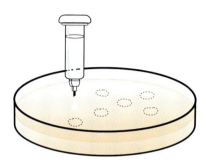

(b) 31 different bacteriophage suspensions are deposited in a fixed pattern.

Phage typing
Figure 10.12

Strain Number	1	2	3	4	5	6	7
1	100						
2	5	100					
3	10	95	100				
4	0	90	95	100			
5	80	15	35	15	100		
6	70	25	40	10	80	100	
7	95	10	20	10	90	75	100

(a)

		A				B		
Strain Number	1	7	5	6	3	2	4
A 1	100						
7	95	100					
5	80	90	100				
6	70	75	80	100			
B 3	10	20	35	40	100		
2	5	10	15	25	95	100	
4	0	10	15	10	95	90	100

(b)

Numerical Taxonomy
Figure 10.15

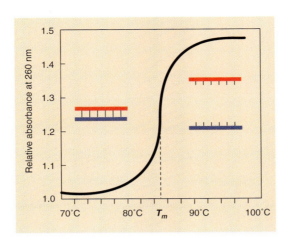

A DNA Melting Curve
Figure 10.16

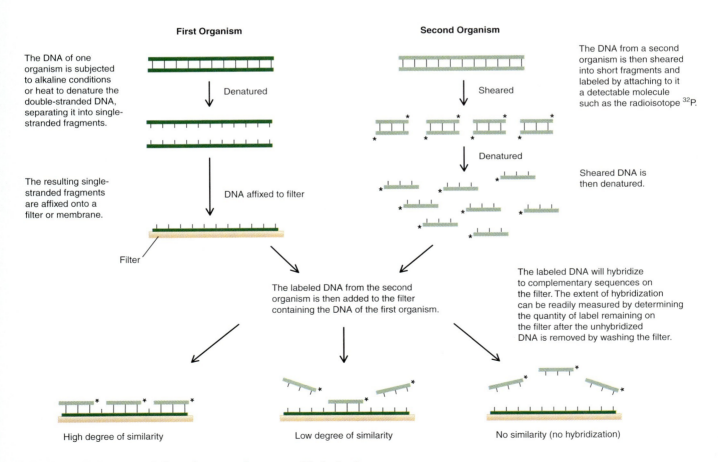

Using DNA Hybridization to Assess Relatedness
Figure 10.17

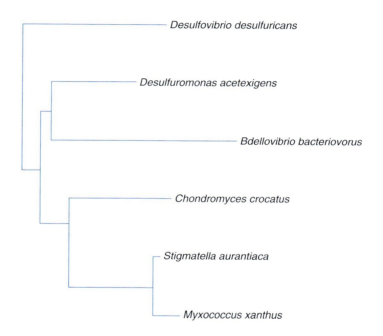

A Phylogenetic Tree
Figure 10.18

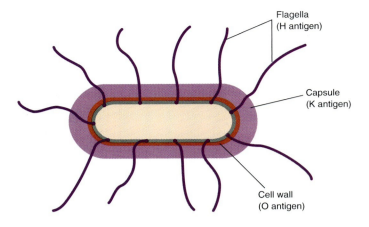

Schematic Drawing of a Member of the Family *Enterobacteriaceae*
Figure 11.15

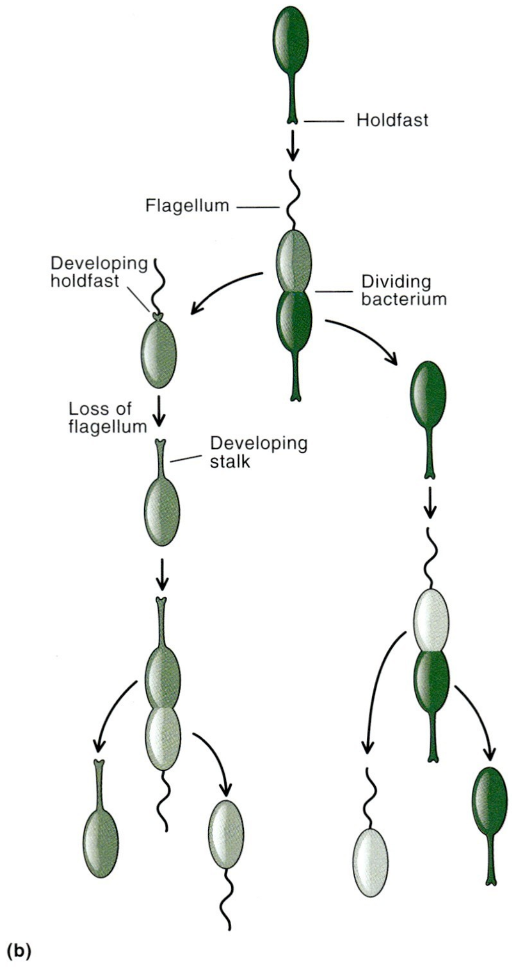

Caulobacter
Figure 11.23

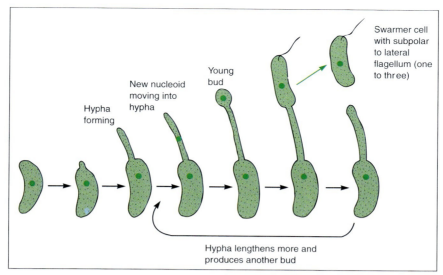

Hyphomicrobium
Figure 11.24

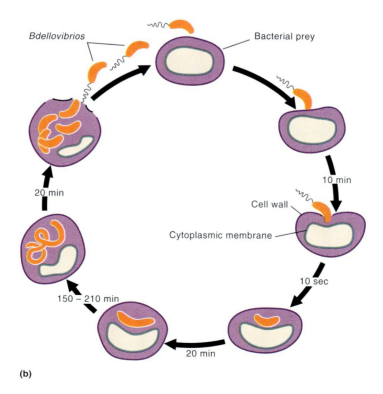

Bdellovibrio
Figure 11.25

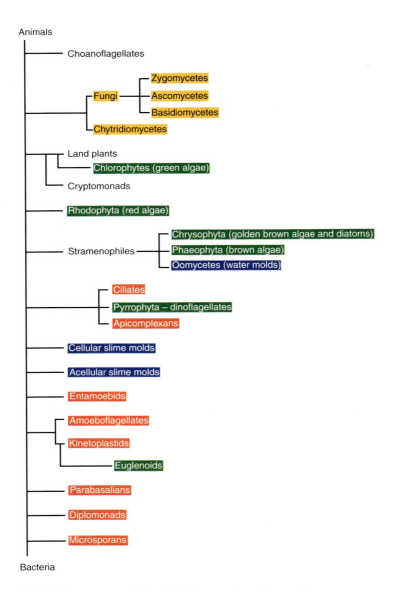

A Phylogeny of the Eukaryotes Based on Ribosomal RNA Sequence Comparison
Figure 12.1

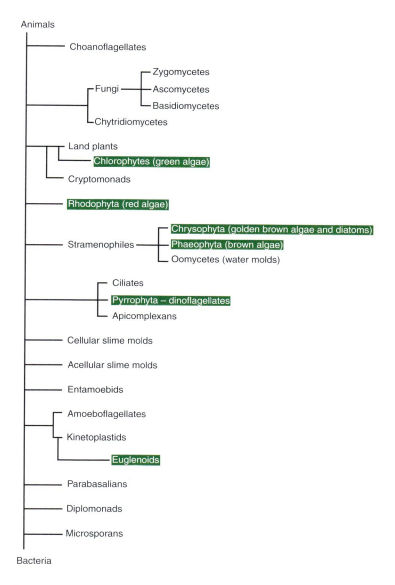

Phylogeny of Algae
Figure 12.3

Binary fission

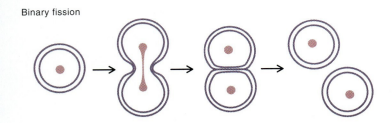

Fragmentation

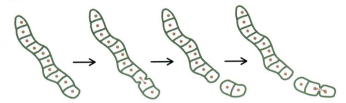

Binary Fission Is an Asexual Reproduction Process in Which a Single Cell Divides into Two Independent Daughter Cells
Figure 12.5

A Phylogenetic Scheme of Eukaryotes Based on rRNA Sequence Comparisons
Figure 12.7

Animals
— Choanoflagellates
— Fungi
  — Zygomycetes
  — Ascomycetes
  — Basidiomycetes
— Chytridiomycetes
— Land plants
— Chlorophytes (green algae)
— Cryptomonads
— Rhodophyta (red algae)
— Stramenophiles
  — Chrysophyta (golden brown algae and diatoms)
  — Phaeophyta (brown algae)
  — Oomycetes (water molds)
— Ciliates
— Pyrrophyta – dinoflagellates
— Apicomplexans
— Cellular slime molds
— Acellular slime molds
— Entamoebids
— Amoeboflagellates
— Kinetoplastids
— Euglenoids
— Parabasalians
— Diplomonads
— Microsporans
Bacteria

110

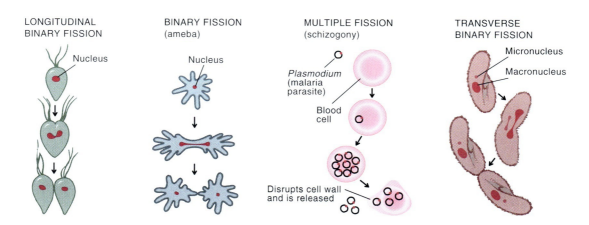

Various Forms of Asexual Reproduction in Protozoa
Figure 12.10

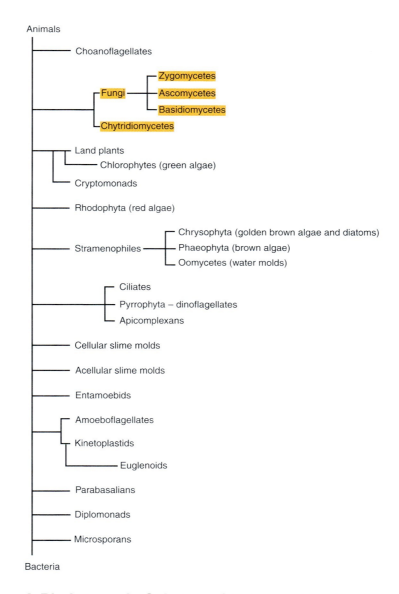

A Phylogenetic Scheme of Eukaryotes Based on rRNA Sequence Comparisons
Figure 12.11

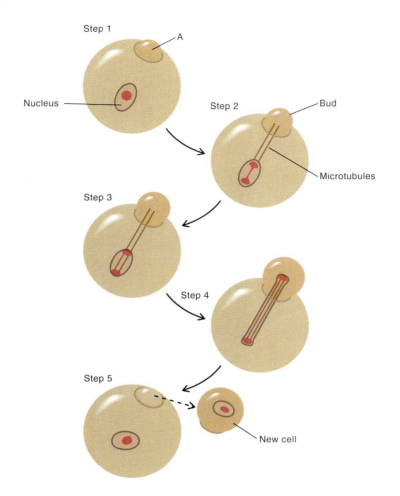

Budding in Yeast
Figure 12.14

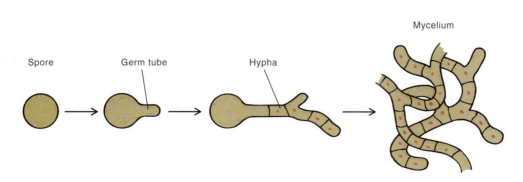

Formation of Hyphae and Mycelium
Figure 12.15

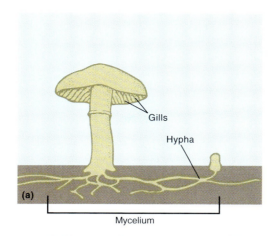

A Meadow Mushroom, *Agaricus campestri*
Figure 12.17

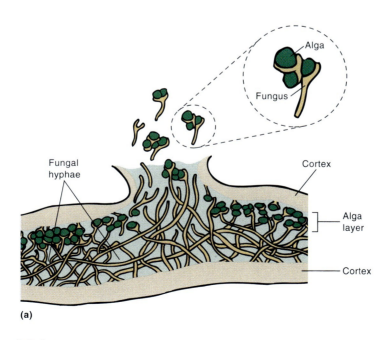

Lichens
Figure 12.18

A Phylogenetic Scheme of Eukaryotes Based on rRNA Sequence Comparisons
Figure 12.19

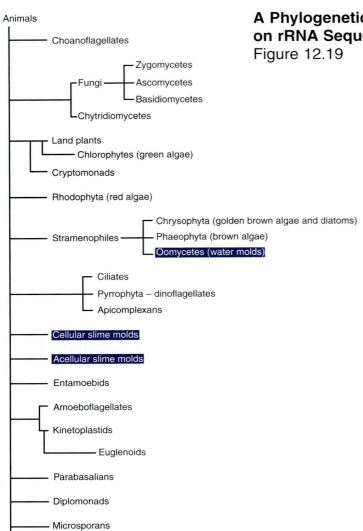

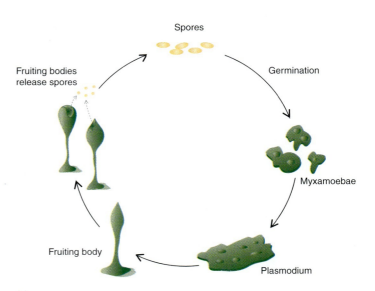

Slime Molds
Figure 12.20 (a)

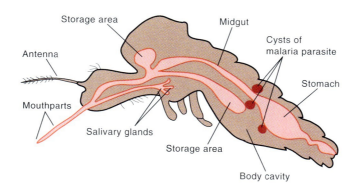

Internal Anatomy of a Mosquito
Figure 12.21

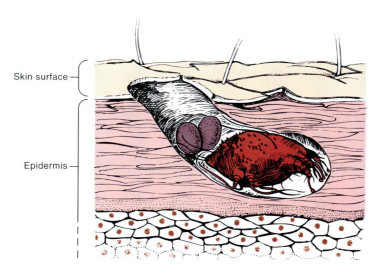

Sarcoptes scabiei **(Scabies Mite)**
Figure 12.23

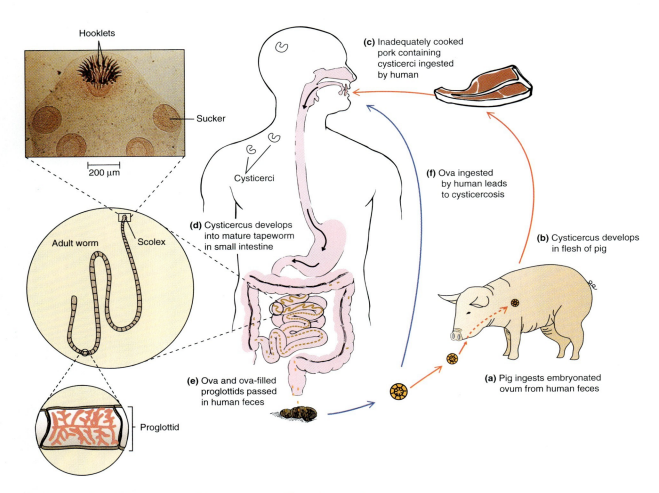

Life Cycle of the Pork Tapeworm, *Taenia solium*, Acquired by Eating Inadequately Cooked Pork
Figure 12.24
Courtesy of S. Eng and F. Schoenknecht

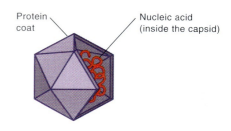
(a) Isometric (adenovirus)

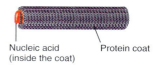
(b) Helical (tobacco mosaic virus)

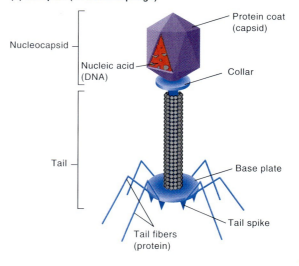
(c) Complex (T4 bacteriophage)

Common Shapes of Viruses
Figure 13.1

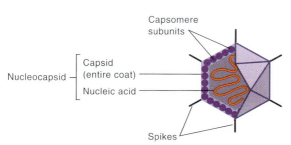
(a) Naked virus

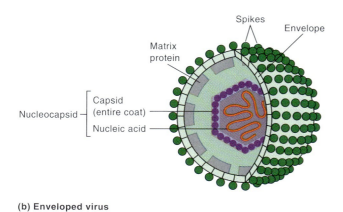
(b) Enveloped virus

Two Different Types of Virions
Figure 13.2

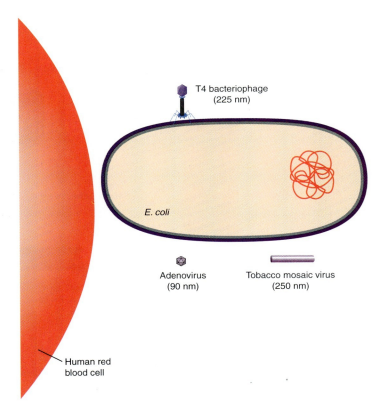

Virion Size
Figure 13.3

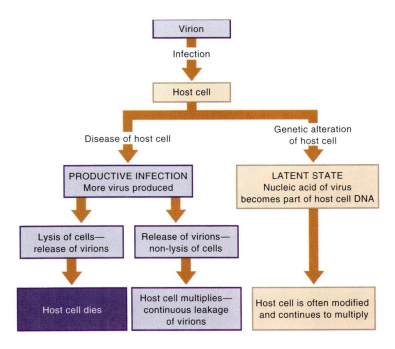

Major Types of Relationships Between Viruses and the Host Cells They Infect
Figure 13.4

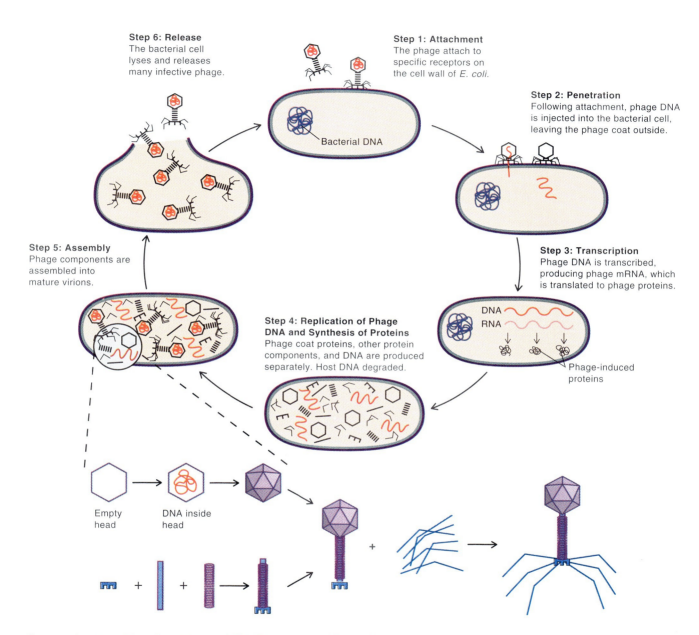

Steps in the Replication of T4 Phage in *E. coli*
Figure 13.5

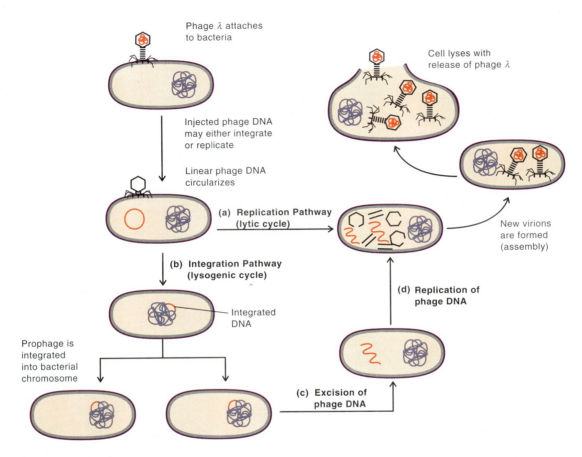

Lambda Phage (λ) Replication Cycle
Figure 13.6

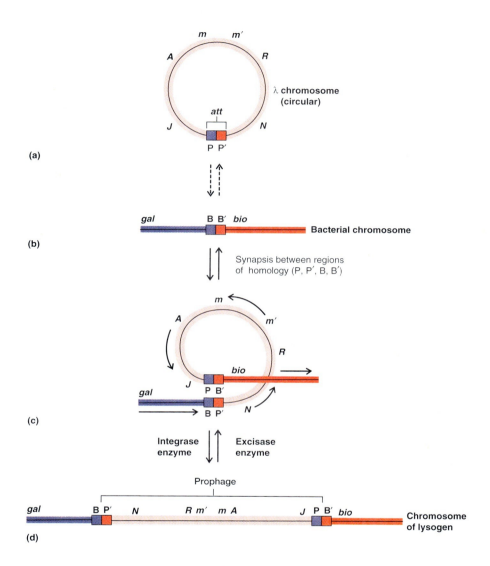

Reversible Insertion and Excision of Lambda (λ) Phage
Figure 13.7

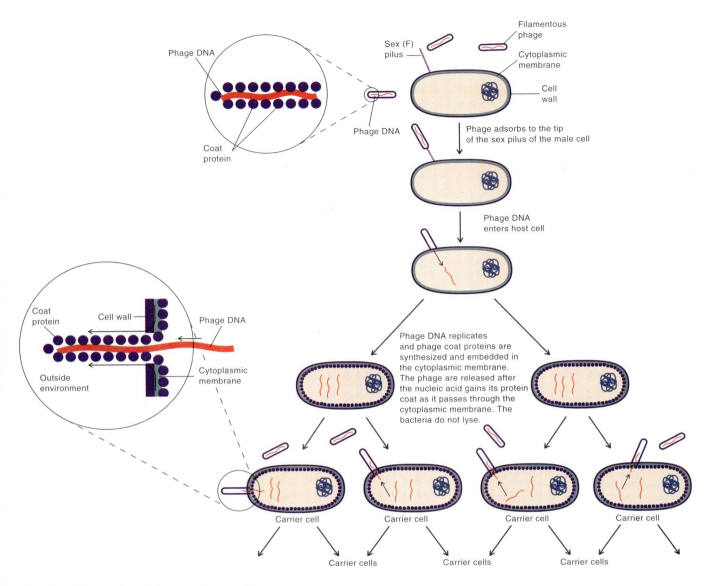

Replication of a Filamentous Phage
Figure 13.9

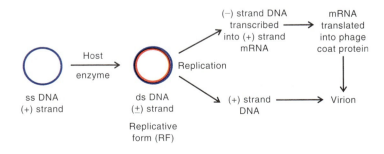

Macromolecule Synthesis in Filamentous Phage Replication
Figure 13.10

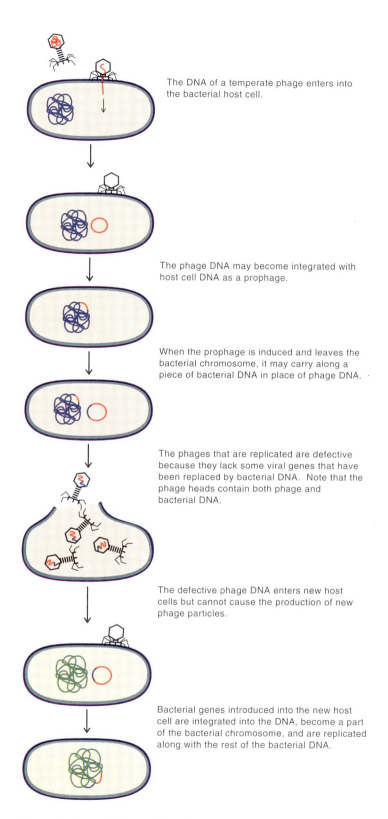

Specialized Transduction by Temperate Phage
Figure 13.11

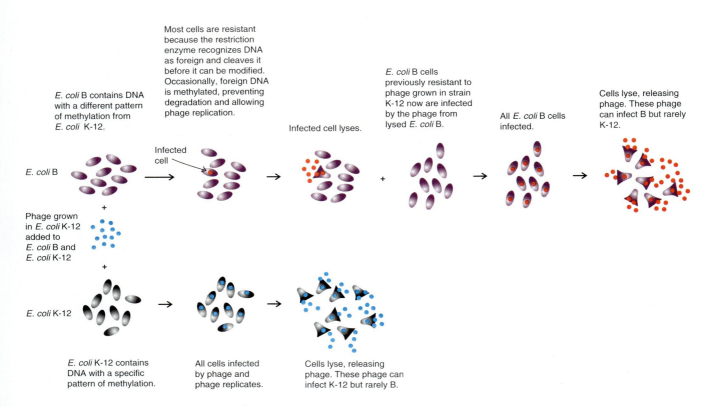

Restriction-Modification System
Figure 13.13

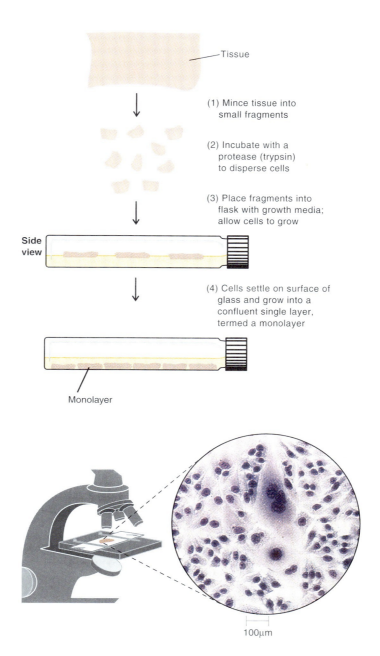

Preparation of Primary Cell Culture
Figure 14.2
© Michael Gabridge/Custom Medical Stock Photo

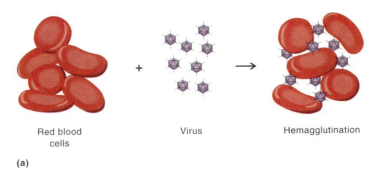

Hemagglutination
Figure 14.6

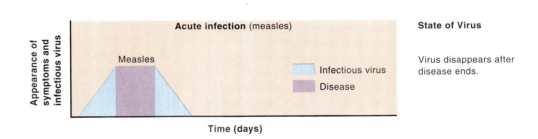

Time Course of Appearance of Symptoms of Measles and the Measles Virions
Figure 14.7

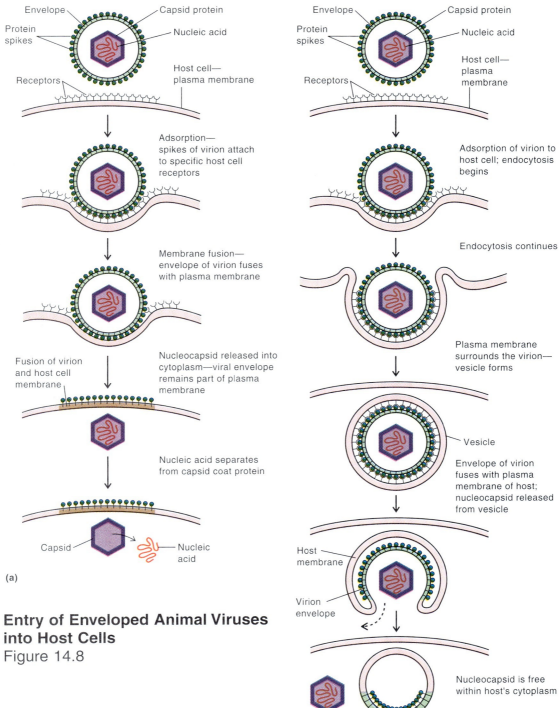

Entry of Enveloped Animal Viruses into Host Cells
Figure 14.8

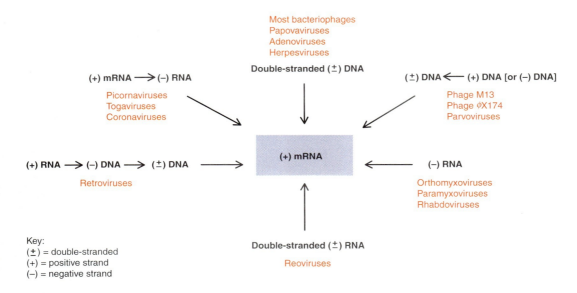

Strategies of Transcription Employed by Different Viruses
Figure 14.9

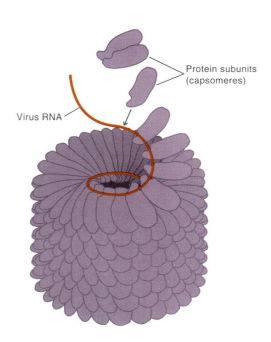

Tobacco Mosaic Virus Assembly
Figure 14.10

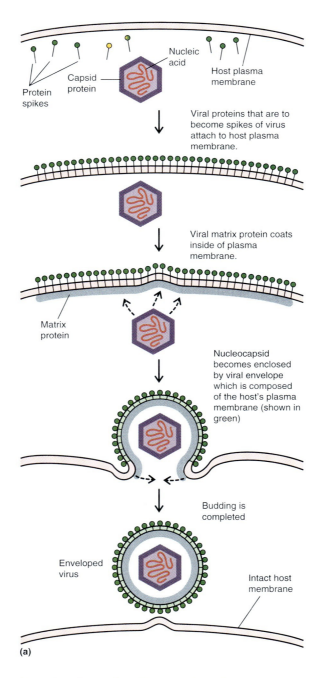

Mechanism for Releasing Enveloped Virions
Figure 14.11

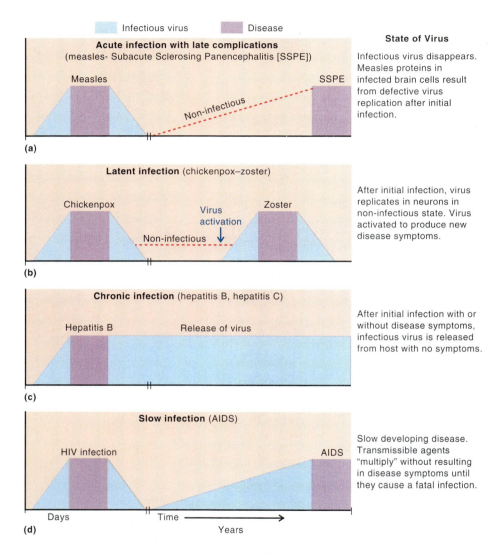

Time Course of Appearance of Disease Symptoms and Infectious Virions in Various Kinds of Viral Infections
Figure 14.12

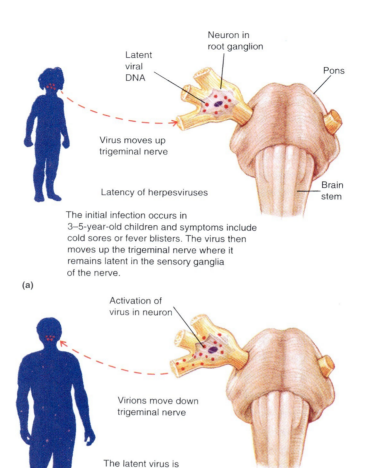

Infection Cycle of Herpes Simplex Virus, HSV-1
Figure 14.13

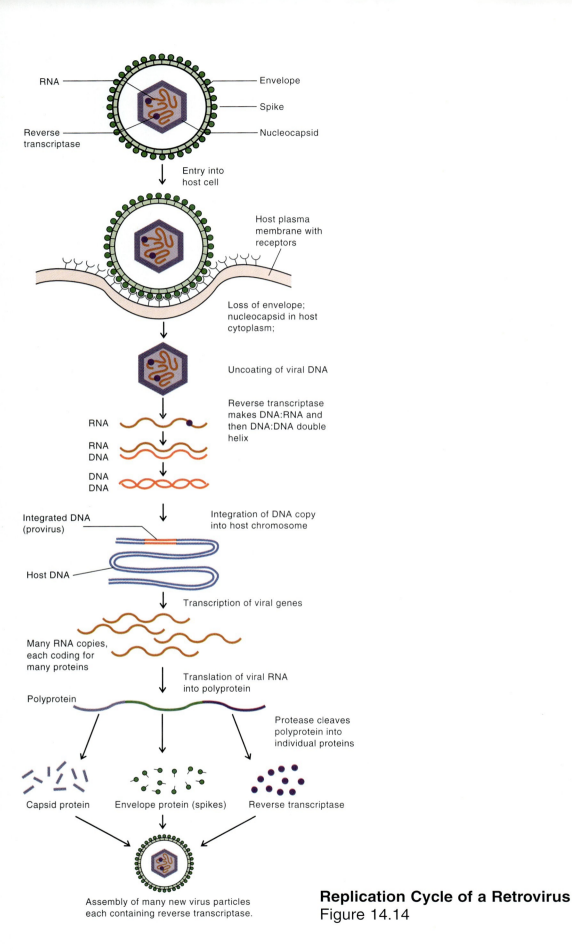

Replication Cycle of a Retrovirus
Figure 14.14

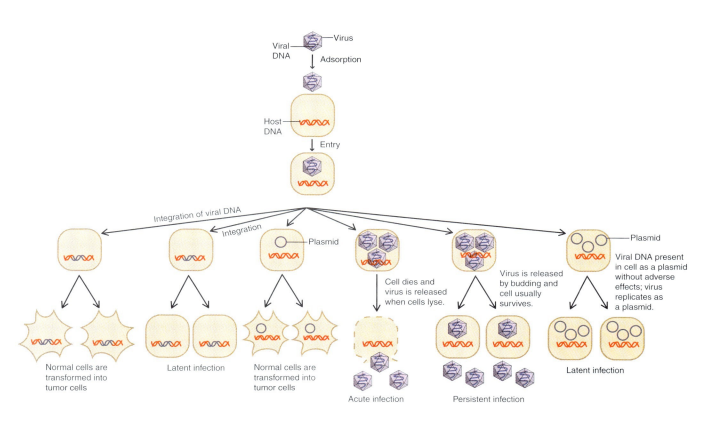

Various Effects of Animal Viruses on the Cells They Infect
Figure 14.15

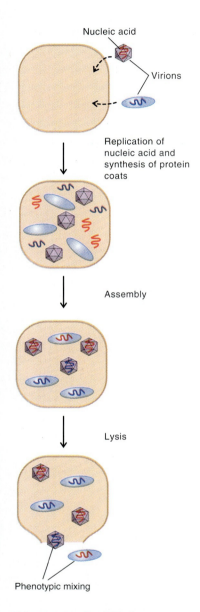

Phenotypic Mixing
Figure 14.16

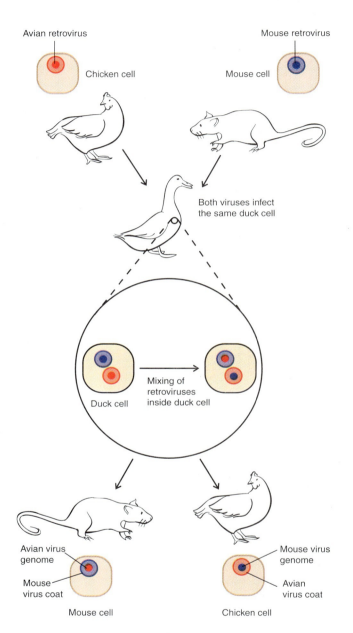

Phenotypic Mixing of Two Retroviruses
Figure 14.17

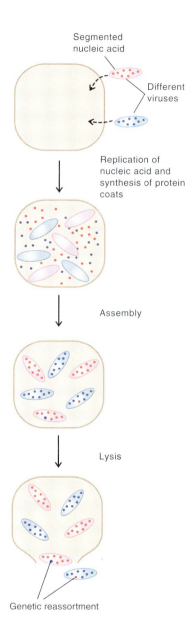

Genetic Reassortment
Figure 14.18

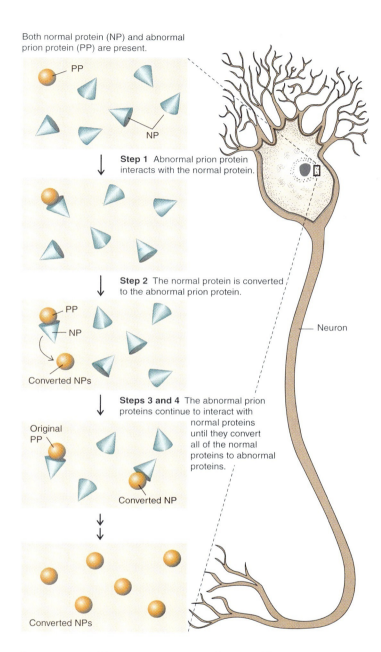

Proposed Mechanism by Which Prions Replicate
Figure 14.22

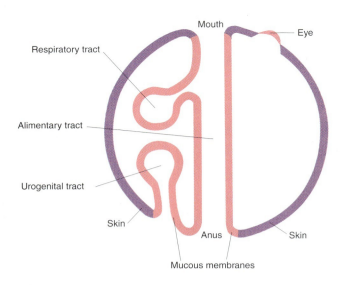

Anatomical Barriers
Figure 15.1

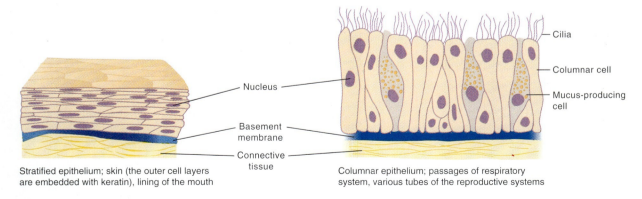

Epithelial Barriers
Figure 15.2

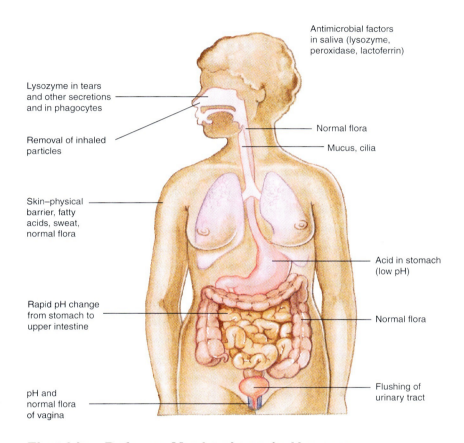

First-Line Defense Mechanisms in Humans
Figure 15.3

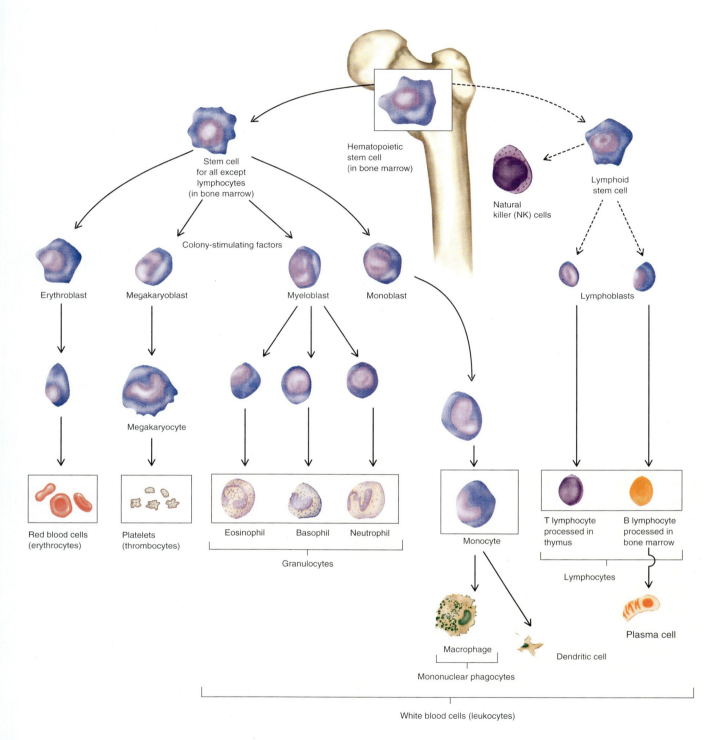

Blood and Lymphoid Cells
Figure 15.4

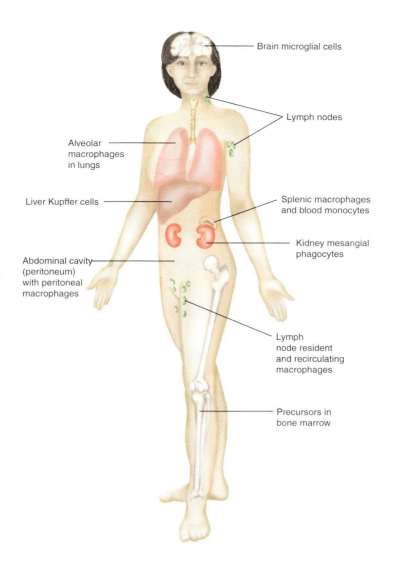

Mononuclear Phagocyte System
Figure 15.5

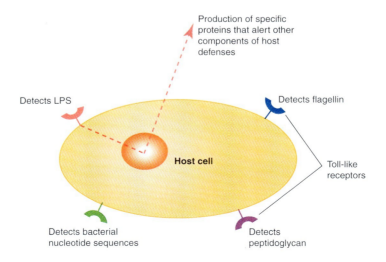

Toll-Like Receptors
Figure 15.6

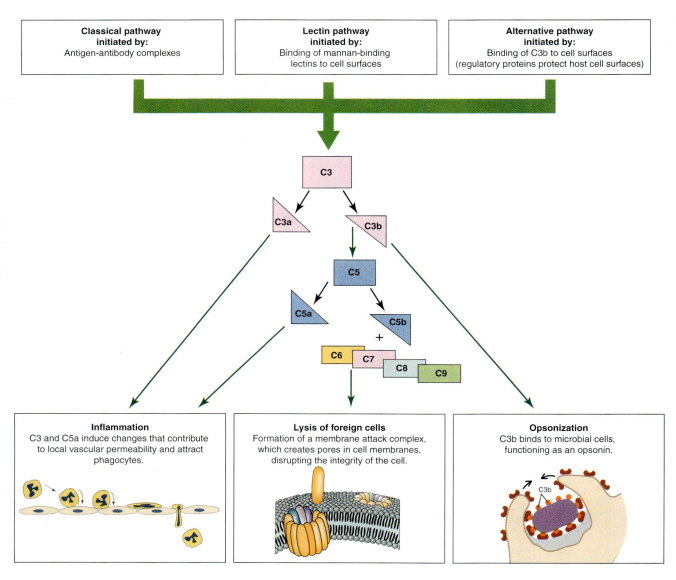

Complement System
Figure 15.7

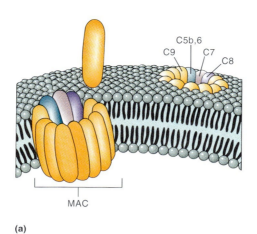

Membrane Attack Complex of Complement (MAC)
Figure 15.8

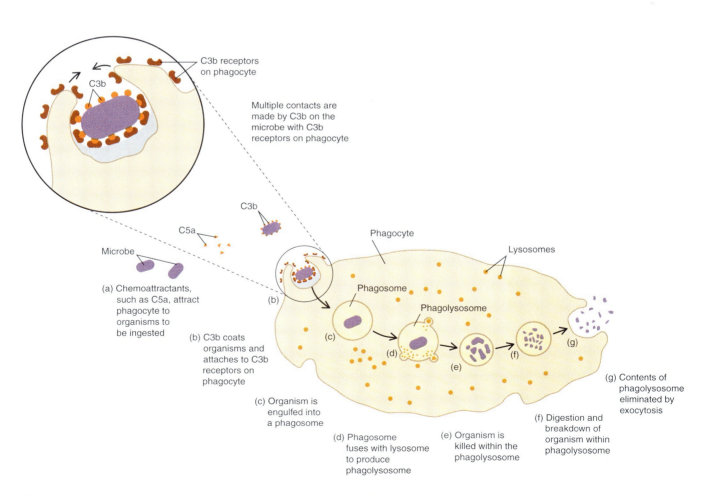

Phagocytosis and Intracellular Destruction of Phagocytized Material
Figure 15.9

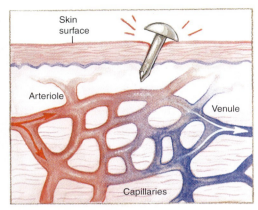

(a) Normal blood flow in the tissues as injury occurs.

- Microbial products
- Microbes
- Tissue damage

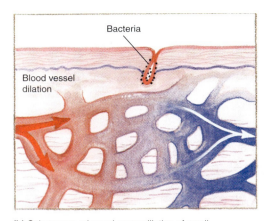

(b) Substances released cause dilation of small blood vessels and increased blood flow in the immediate area.

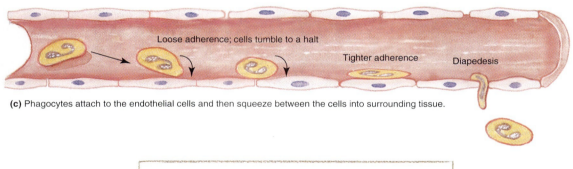

(c) Phagocytes attach to the endothelial cells and then squeeze between the cells into surrounding tissue.

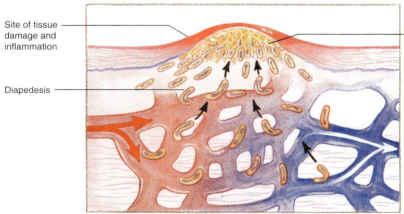

(d) The attraction of phagocytes causes them to move to the site of damage and inflammation. Collections of dead phagocytes and tissue debris make up the pus often found at sites of an active inflammatory response.

The Inflammatory Process
Figure 15.10

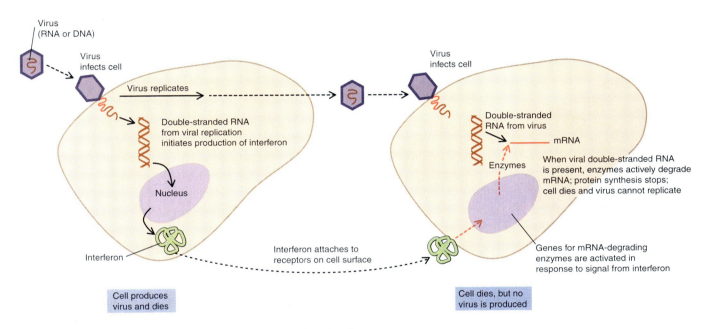

Mechanism of the Antiviral Activity of Interferons
Figure 15.11

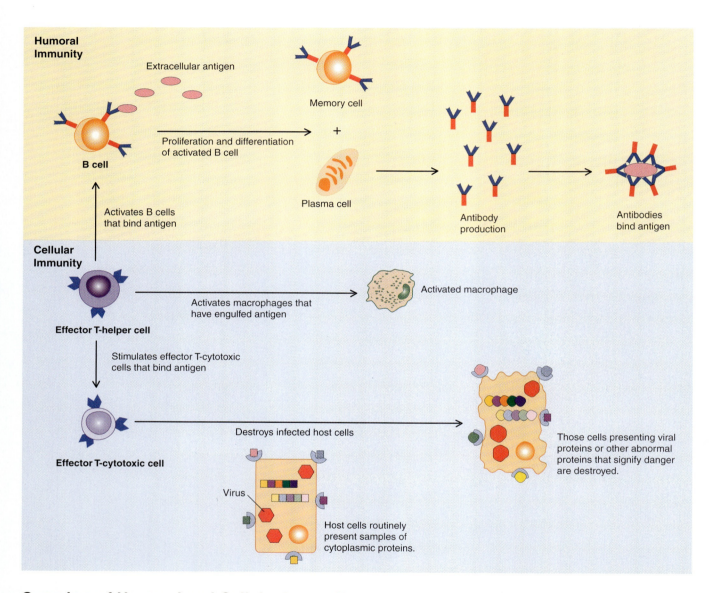

Overview of Humoral and Cellular Immunity
Figure 16.1

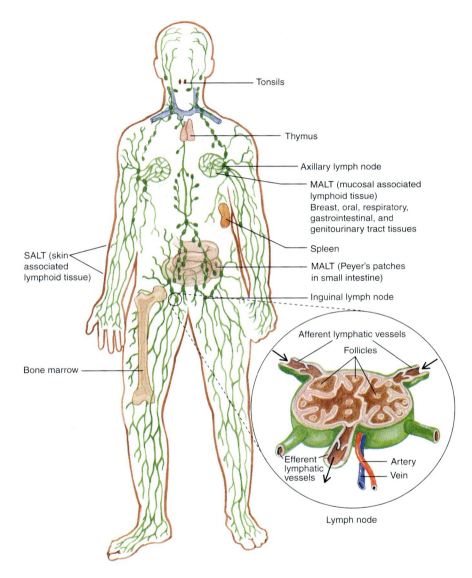

Anatomy of the Lymphoid System
Figure 16.2

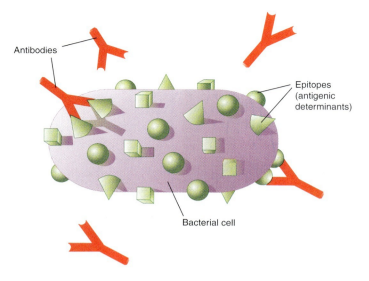

Antibodies and Antigen Epitopes on a Bacterial Cell
Figure 16.3

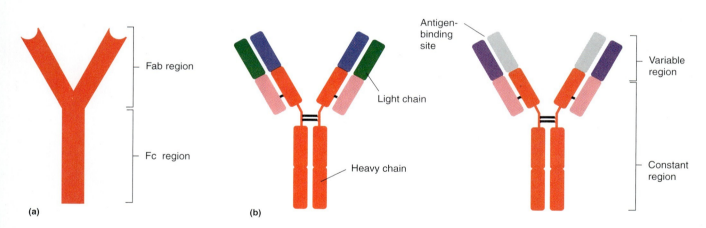

Basic Structure of an Antibody Molecule
Figure 16.4

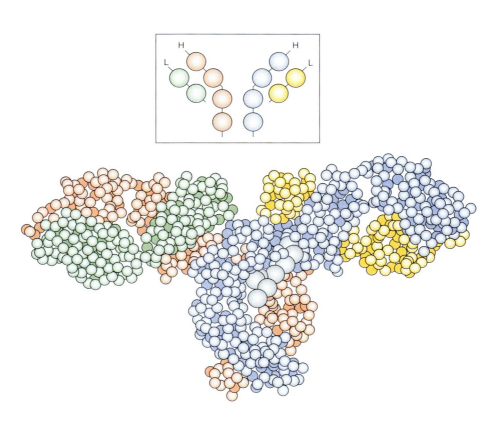

Model of an IgG Molecule
Figure 16.5

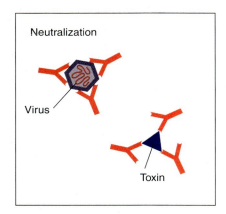
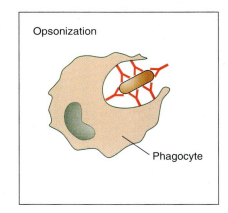
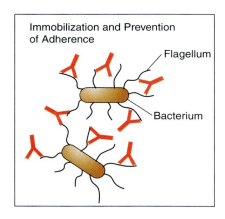
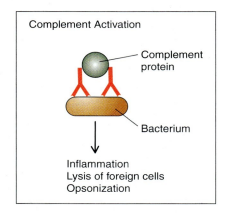
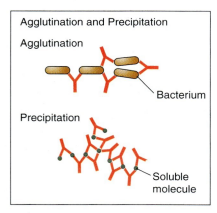
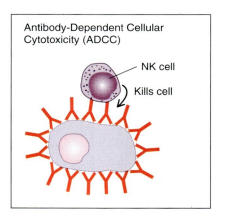

Protective Outcomes of Antibody-Antigen Binding
Figure 16.6

Immunoglobulin G Levels in the Fetus and Infant
Figure 16.7

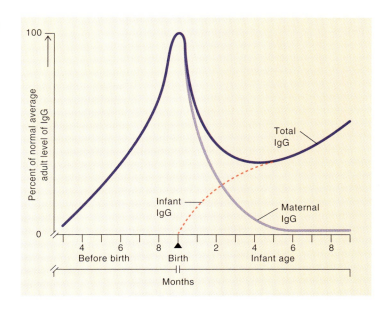

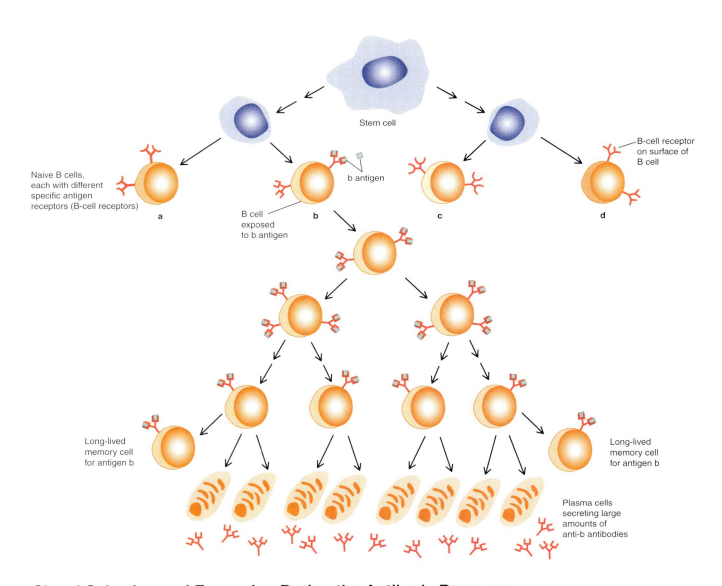

Clonal Selection and Expansion During the Antibody Response
Figure 16.8

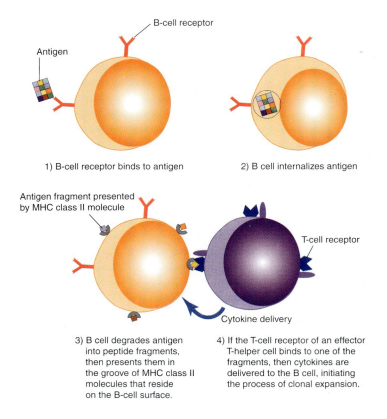

Antigen Presentation by a B Cell
Figure 16.9

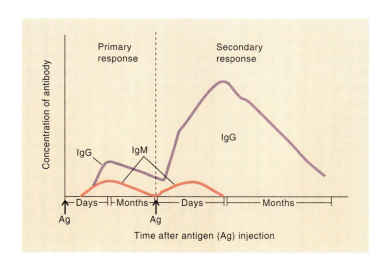

The Primary and Secondary Responses to Antigen
Figure 16.11

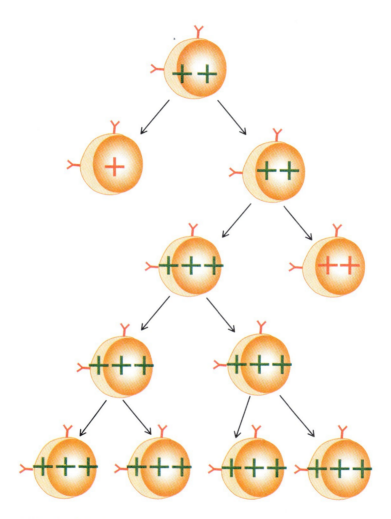

Affinity Maturation
Figure 16.12

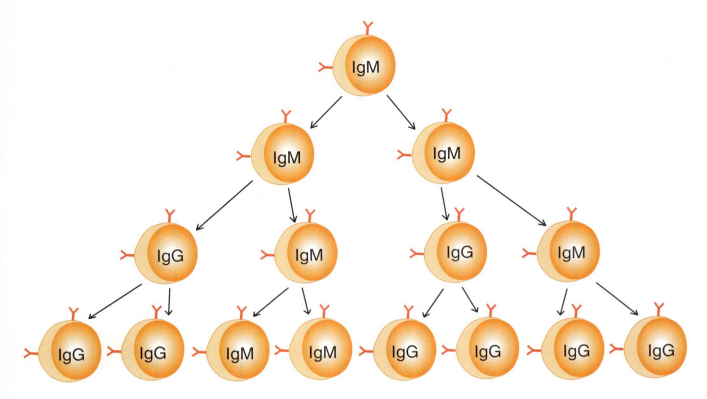

Class Switching
Figure 16.13

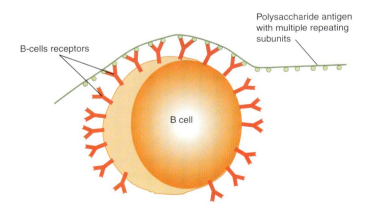

T-Independent Antigens
Figure 16.14

Two T-Cell Receptors Composed of Alpha and Beta Polypeptide Chains
Figure 16.15

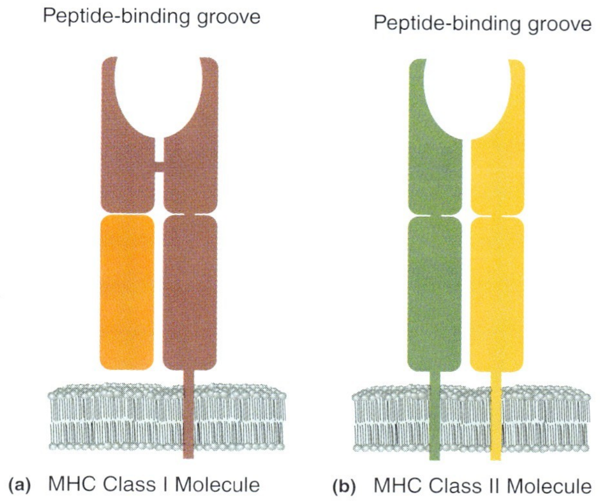

MHC Molecules
Figure 16.16

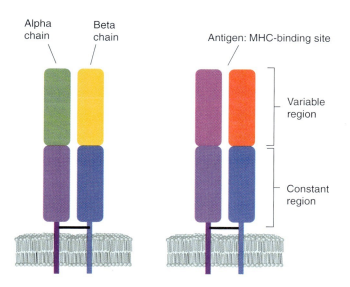

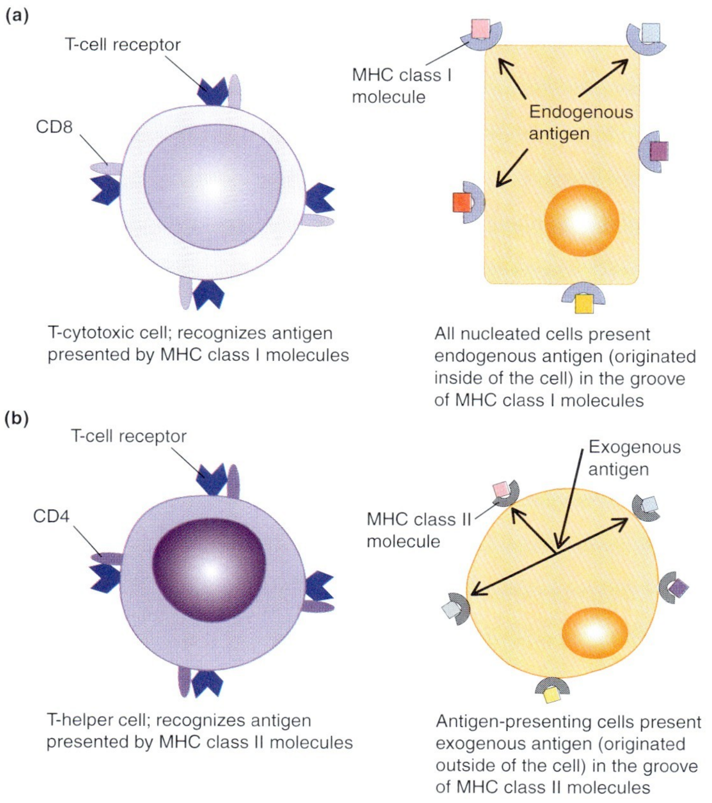

Antigen Recognition by T Cells
Figure 16.17

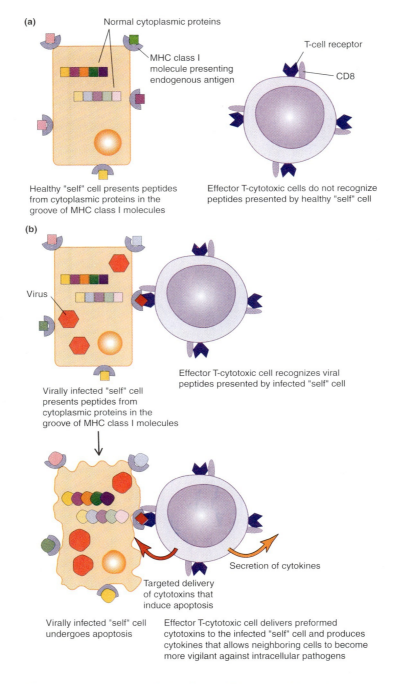

Consequences of Antigen Recognition by Effector T-Cytotoxic Cells
Figure 16.18

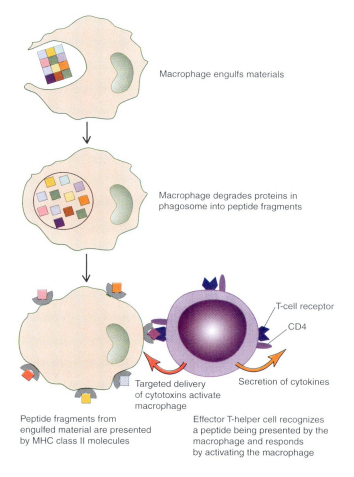

Antigen Presentation by a Macrophage to an Effector T-Helper Cell
Figure 16.19

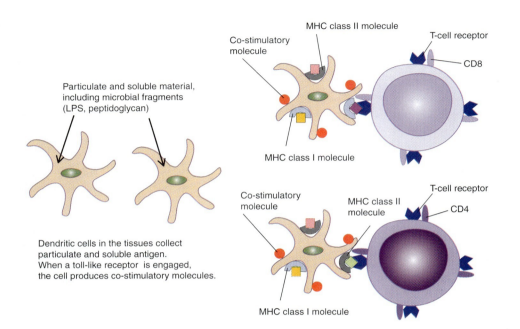

Activation of T Cells by Dendritic Cells Expressing Co-Stimulatory Molecules
Figure 16.20

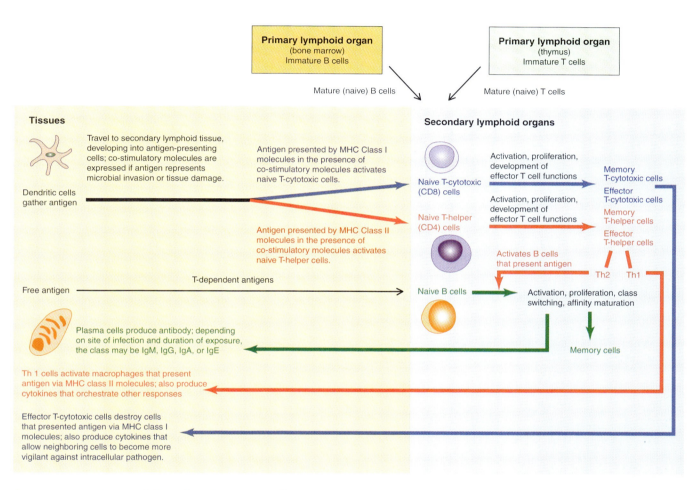

Summary of the Adaptive Immune Response
Figure 16.21

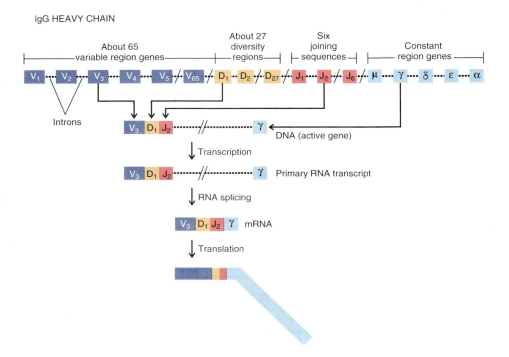

Antibody Diversity
Figure 16.22

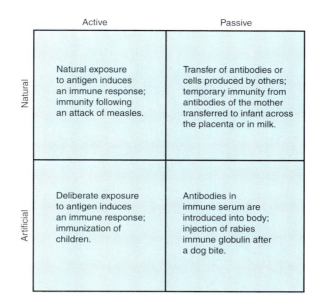

Acquired Immunity
Figure 17.1

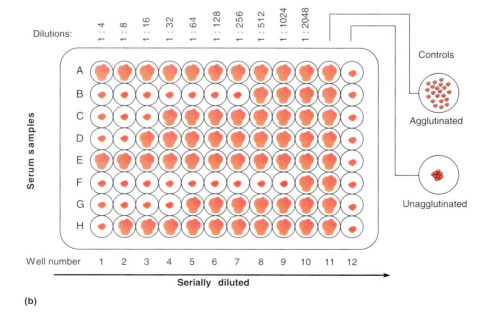

Quantitation of Immunologic Tests
Figure 17.2

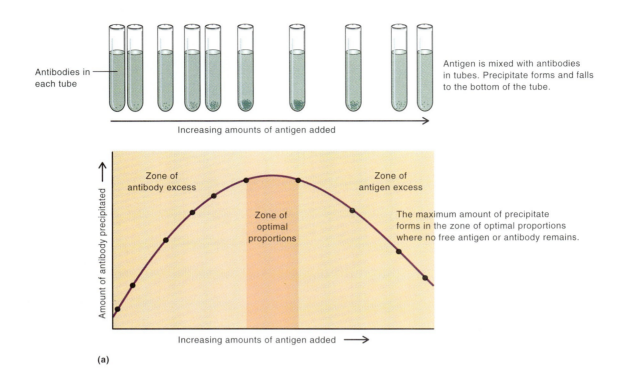

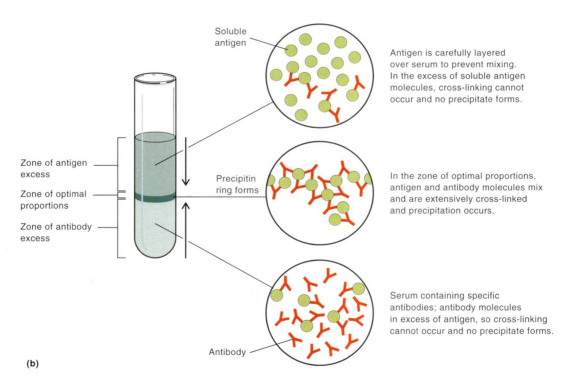

Antigen-Antibody Precipitation Reactions
Figure 17.3

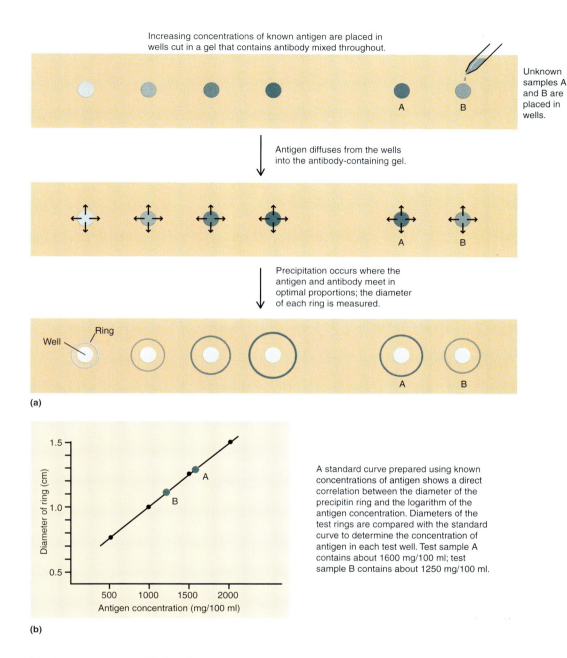

Radial Immunodiffusion Is Used to Measure the Concentration of Antigen in a Sample
Figure 17.4

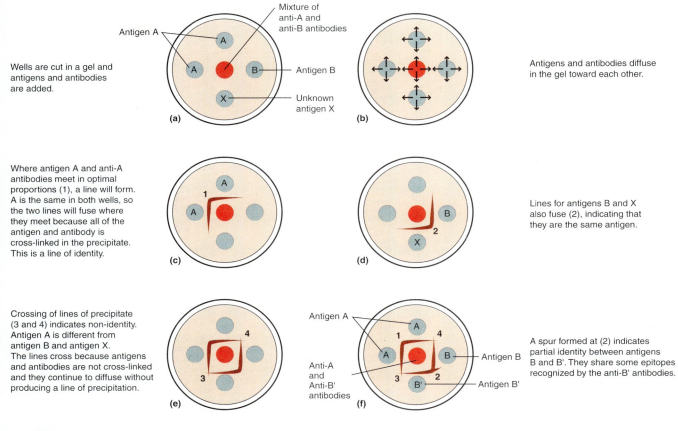

Double Immunodiffusion
Figure 17.5

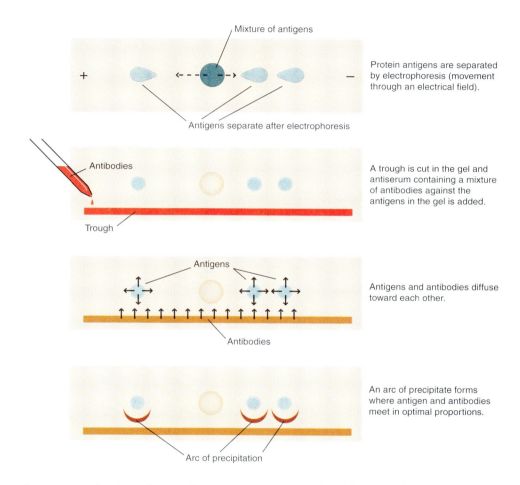

Immunoelectrophoresis Permits Identification of Antigens in a Mixture
Figure 17.6

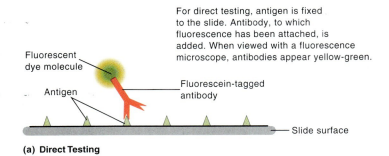
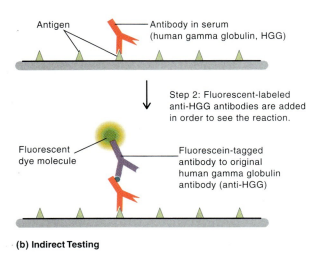

Fluorescent Antibody Tests
Figure 17.8

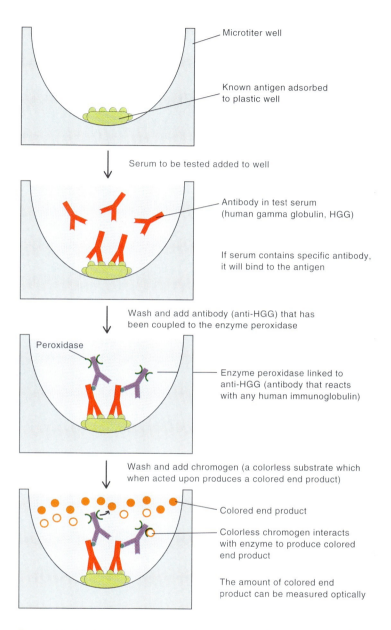

Enzyme-Linked Immunosorbent Assay (ELISA)
Figure 17.9

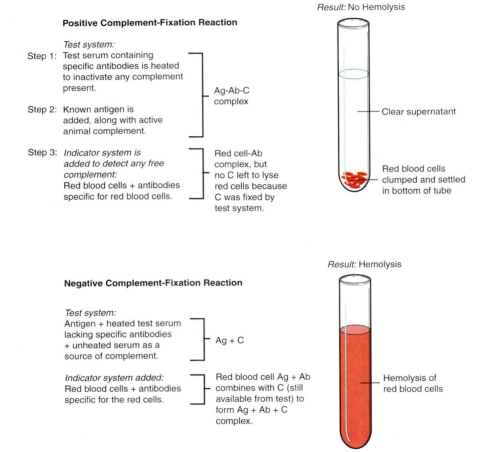

Complement Fixation Test Procedure
Figure 17.12

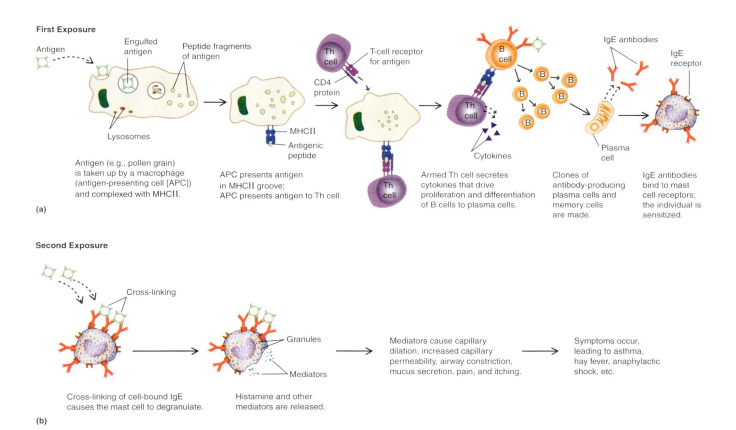

Mechanisms of Type I Hypersensitivity: Immediate IgE-Mediated
Figure 18.1

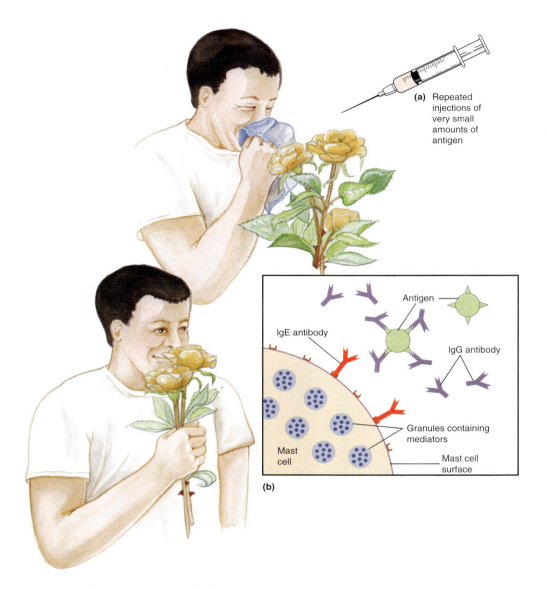

Immunotherapy for IgE Allergies
Figure 18.3

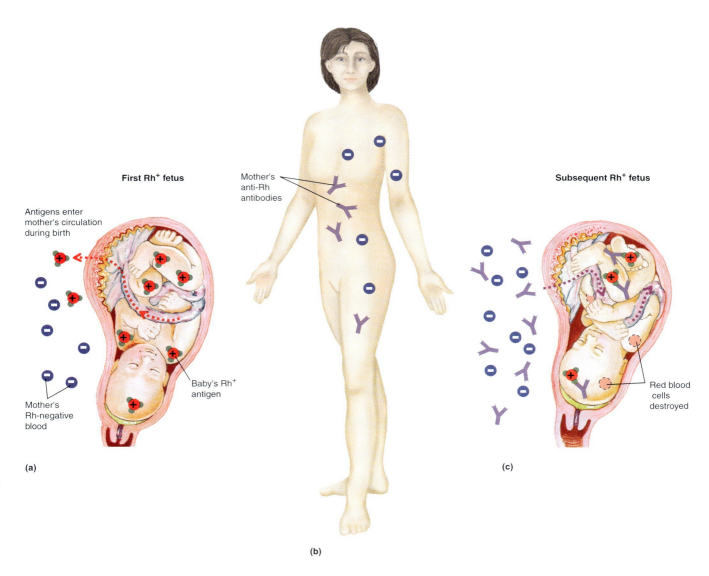

Hemolytic Disease of the Newborn
Figure 18.4

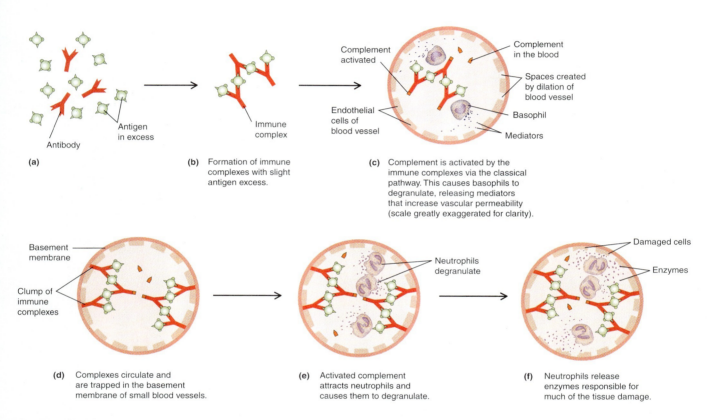

Type III Hypersensitivity: Immune Complex-Mediated
Figure 18.5

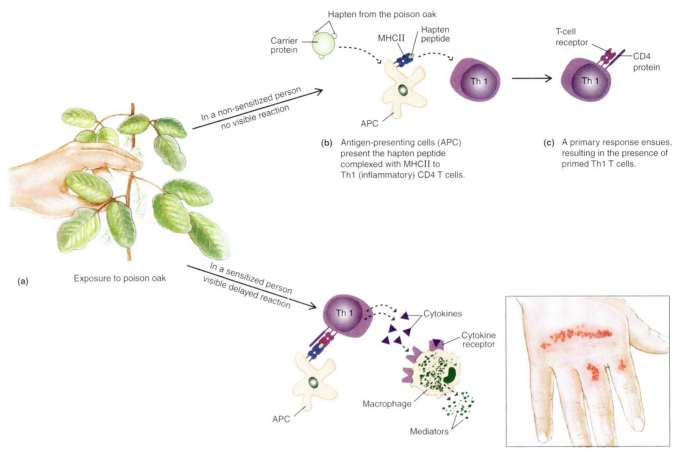

Poison Oak Dermatitis Is an Example of Type IV Hypersensitivity: Delayed Cell-Mediated
Figure 18.7

(a) Exposure to poison oak

In a non-sensitized person no visible reaction

(b) Antigen-presenting cells (APC) present the hapten peptide complexed with MHCII to Th1 (inflammatory) CD4 T cells.

(c) A primary response ensues, resulting in the presence of primed Th1 T cells.

In a sensitized person visible delayed reaction

(d) Antigen-presenting cells present the hapten-peptide complex to sensitized Th1 cells, which secrete cytokines and attract macrophages; the macrophages are activated and secrete mediators of inflammation that cause skin lesions.

(e) Characteristic skin lesions appear after 24 hours, reaching their peak at 48–72 hours after exposure to the plant.

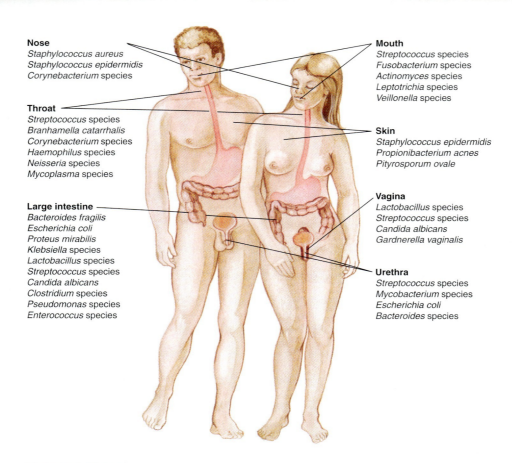

Normal Flora
Figure 19.1

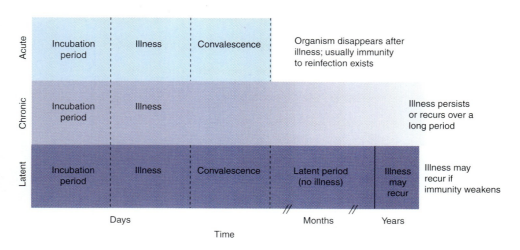

The Course of Infectious Diseases
Figure 19.2

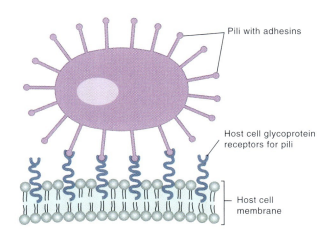

Pili Attachment to Host Cell
Figure 19.3

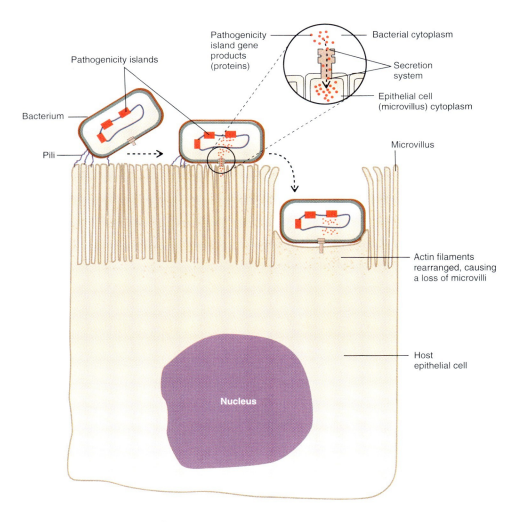

Type III Secretion Systems
Figure 19.4

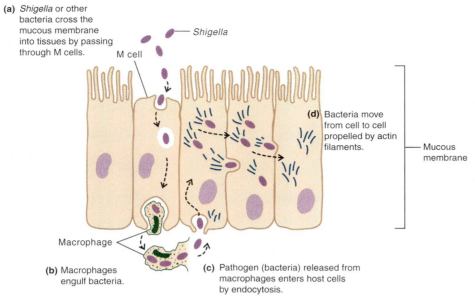

Antigen-Sampling Processes Provide a Mechanism for Invasion
Figure 19.6

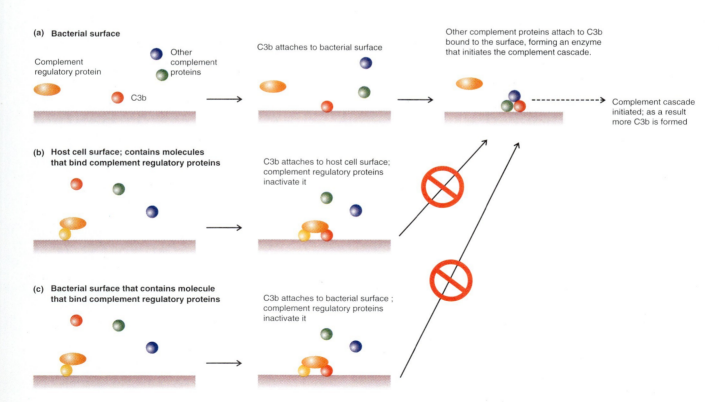

Avoiding the Alternative Pathway of Complement Activation
Figure 19.8

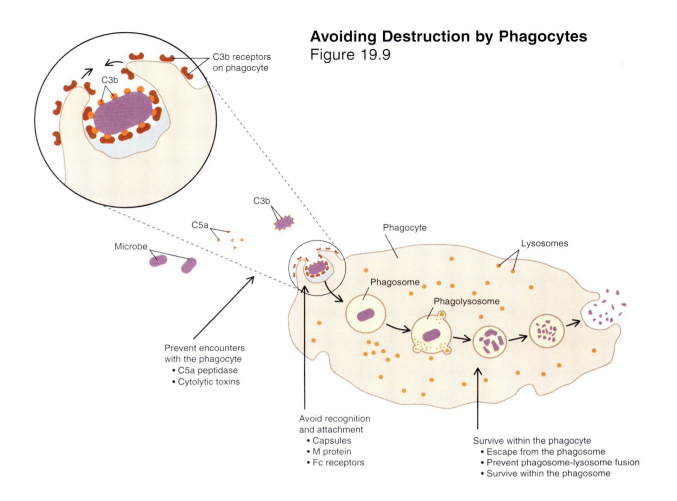

Avoiding Destruction by Phagocytes
Figure 19.9

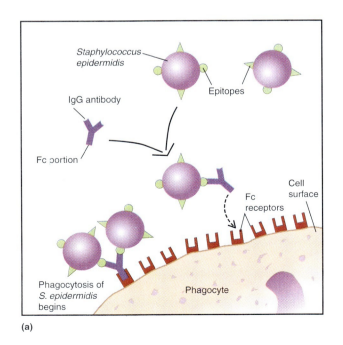

(a)

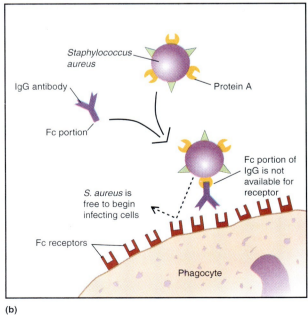

(b)

Foiling Opsonization by Antibodies
Figure 19.11

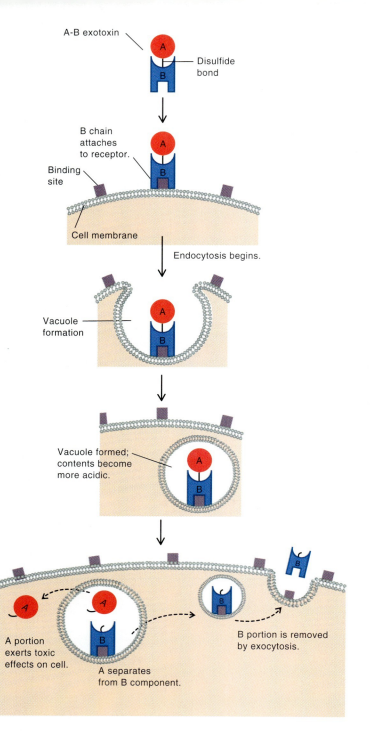

The Action of A-B Exotoxins
Figure 19.12

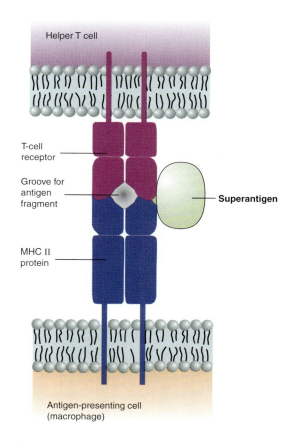

Superantigens
Figure 19.13

Endotoxin
Figure 19.14

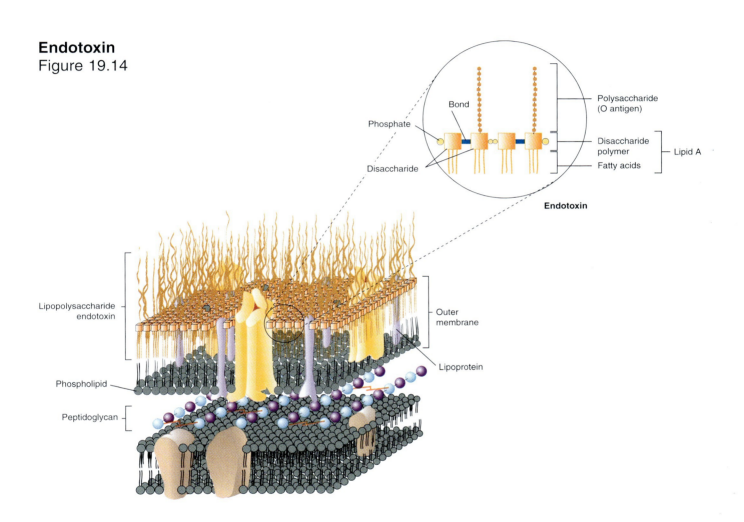

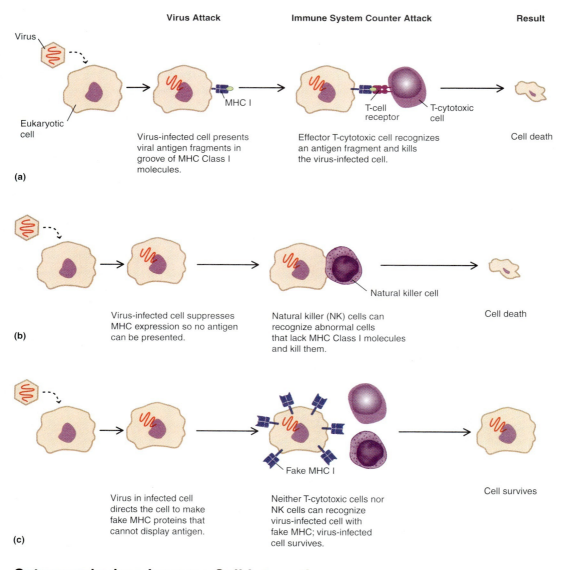

Cytomegalovirus-Immune Cell Interactions
Figure 19.15

Spread of Pathogens
Figure 20.1

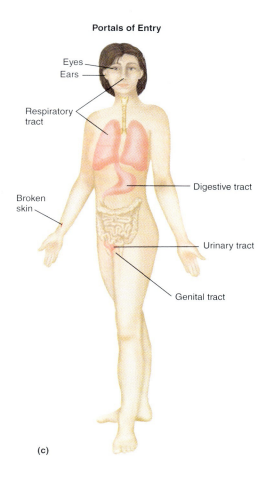

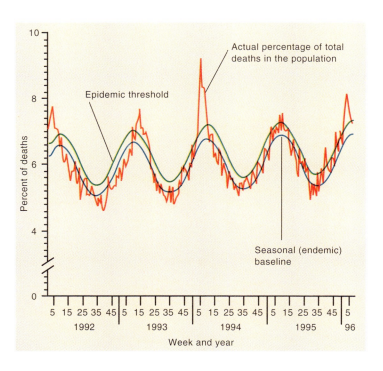

Incidence of Influenza, and Endemic Disease that Can Be Epidemic
Figure 20.2

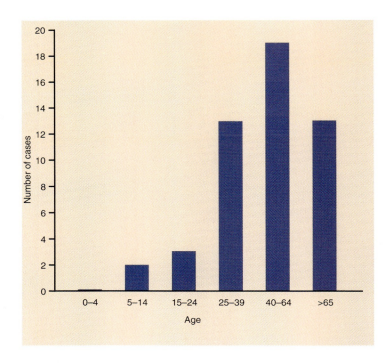

Incidence of Tetanus by Age Group
Figure 20.4

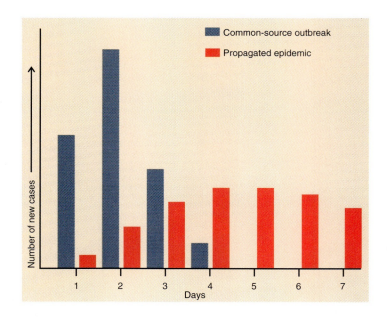

Comparison of Propagated Versus Common Source Epidemics
Figure 20.5

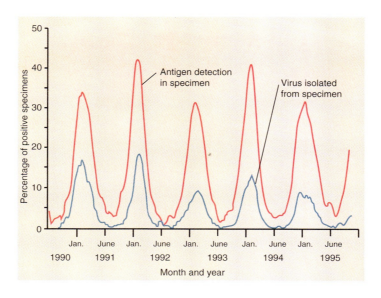

Seasonal Occurrence of Respiratory Infections Caused by Respiratory Syncytial Virus
Figure 20.6

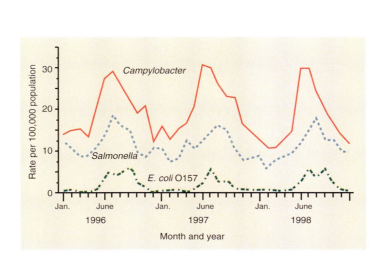

Seasonal Occurrence of Gastrointestinal Diseases
Figure 20.7

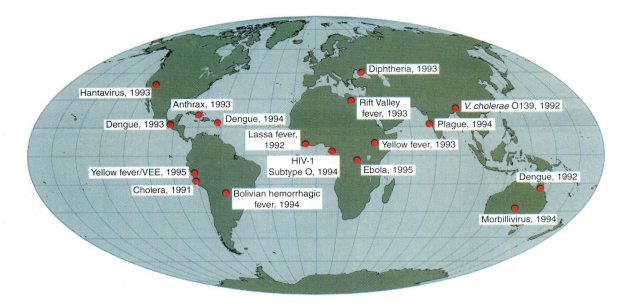

World Map of Emerging Diseases
Figure 20.10

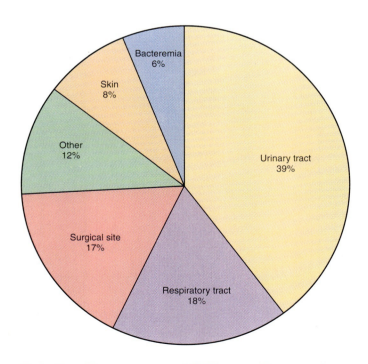

Relative Frequency of Different Types of Nosocomial Infections
Figure 20.11

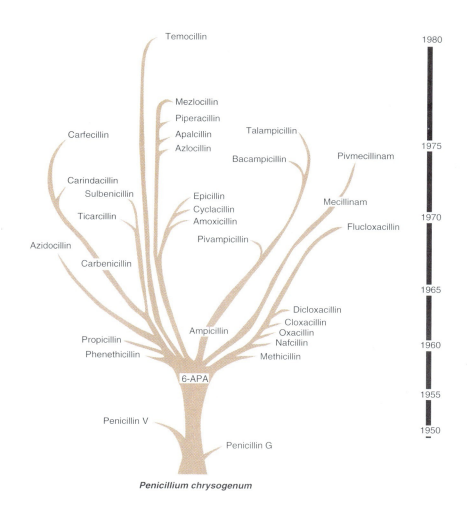

Family Tree of Penicillins
Figure 21.1

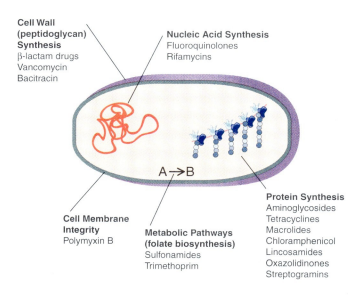

Targets of Antibacterial Medications
Figure 21.2

Antibacterial Medications that Interfere with Cell Wall Biosynthesis
Figure 21.3

β-lactam drugs
Interfere with the formation of the peptide side chains between adjacent strands of peptidoglycan by inhibiting penicillin-binding proteins

Vancomycin
Binds to the amino acid side chain of NAM molecules, interfering with peptidoglycan synthesis

Bacitracin
Interferes with the transport of peptidoglycan precursors across the cytoplasmic membrane

Peptidoglycan (cell wall)
Cytoplasmic membrane

NAG
NAM

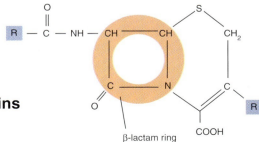

The β-Lactam Ring of Penicillins and Cephalosporins
Figure 21.4

(a) Penicillin
(b) Cephalosporin

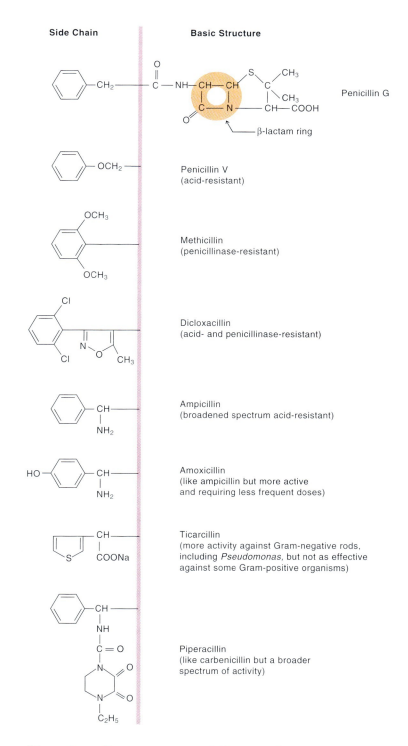

Chemical Structures and Properties of Representative Members of the Penicillin Family
Figure 21.6

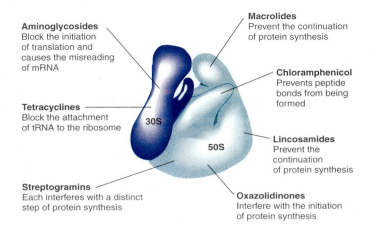

Antibacterial Medications that Inhibit Prokaryotic Protein Synthesis
Figure 21.7

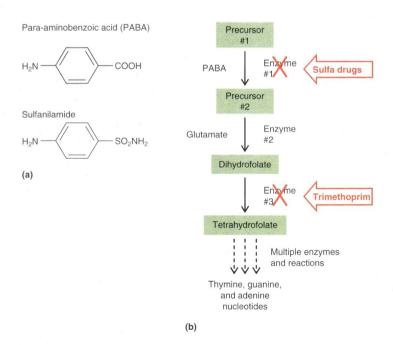

Inhibitors of the Folate Pathway
Figure 21.8

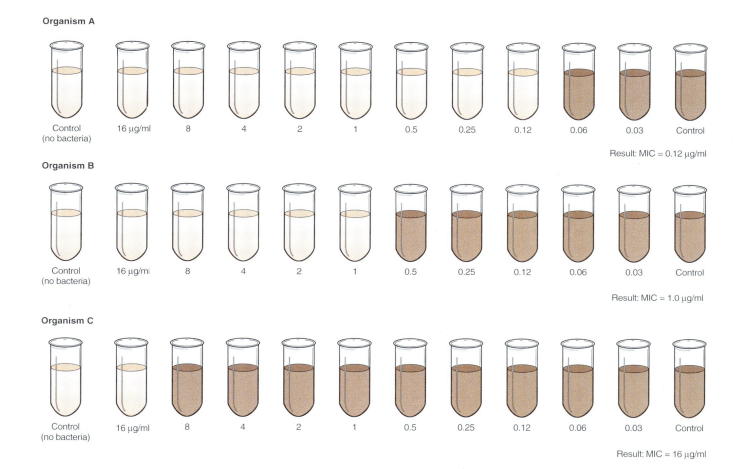

Determining the Minimum Inhibitory Concentration (MIC) of an Antimicrobial Drug
Figure 21.9

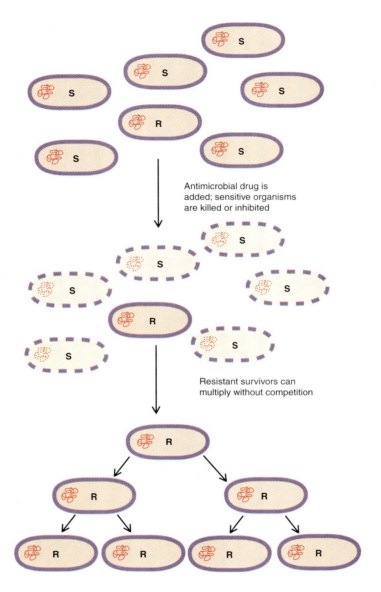

The Selective Advantage of Drug Resistance
Figure 21.13

Common Mechanisms of Antimicrobial Drug Resistance
Figure 21.14

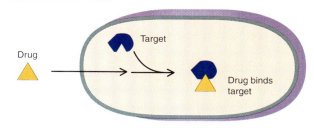

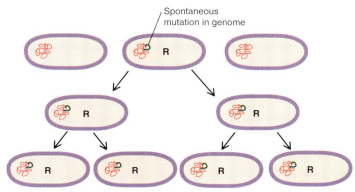

(a) Vertical Evolution

(b) Horizontal Evolution

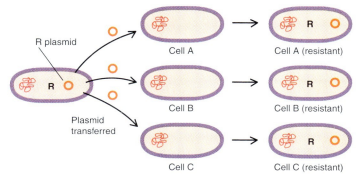

The Acquisition of Antimicrobial Resistance
Figure 21.15

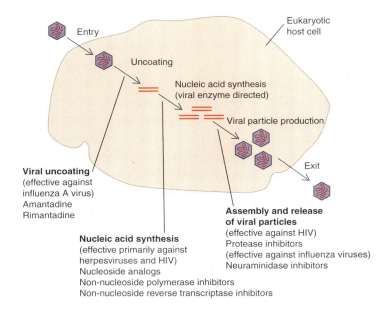

Targets of Antiviral Drugs
Figure 21.16

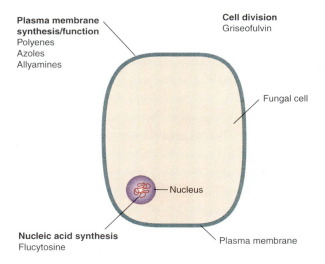

Targets of Antifungal Drugs
Figure 21.17

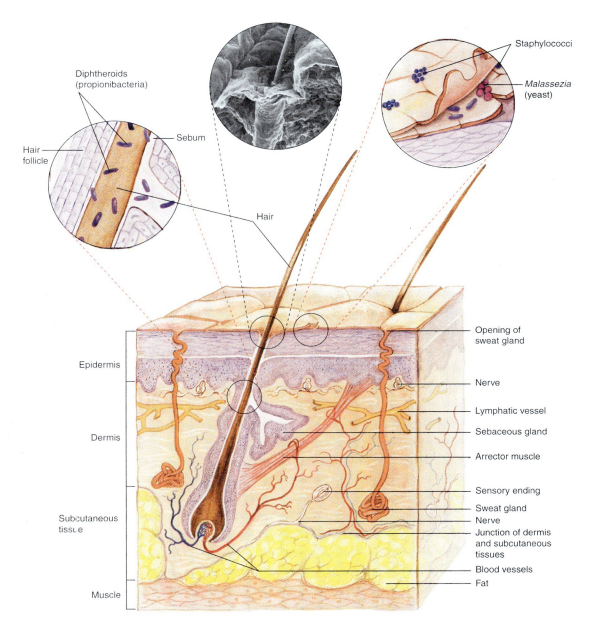

Microscopic Anatomy of the Skin
Figure 22.1
Courtesy Karen A. Holbrook

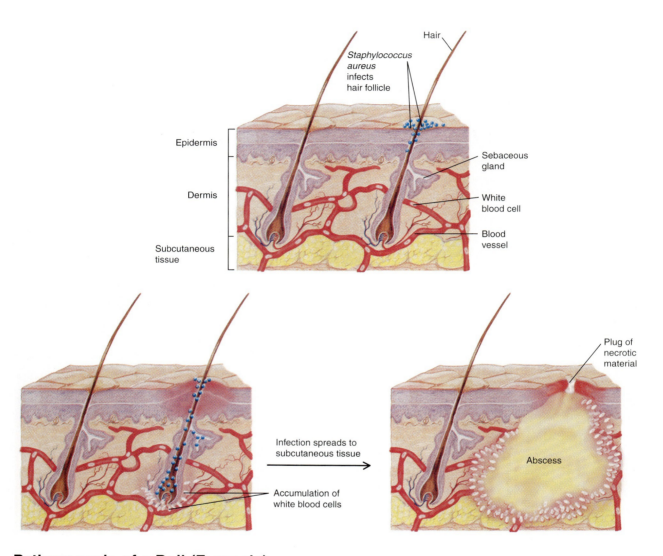

Pathogenesis of a Boil (Furuncle)
Figure 22.3

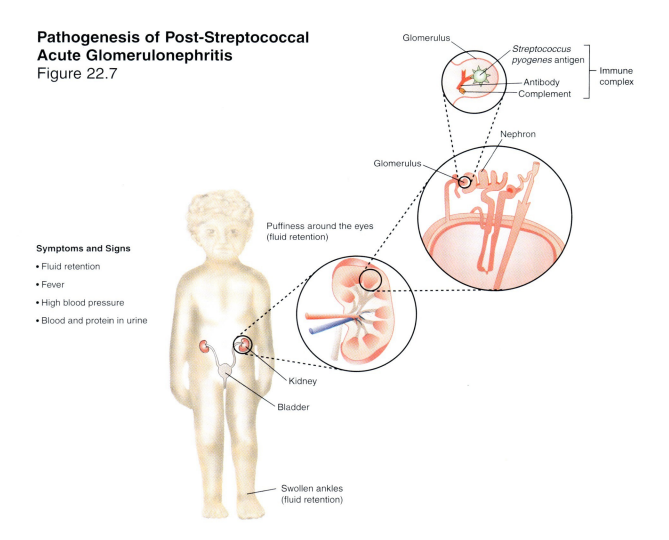

Pathogenesis of Post-Streptococcal Acute Glomerulonephritis
Figure 22.7

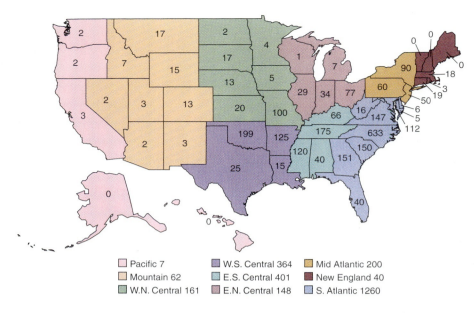

Total Reported Cases of Rocky Mountain Spotted Fever by State and Region, 1994–1998
Figure 22.10

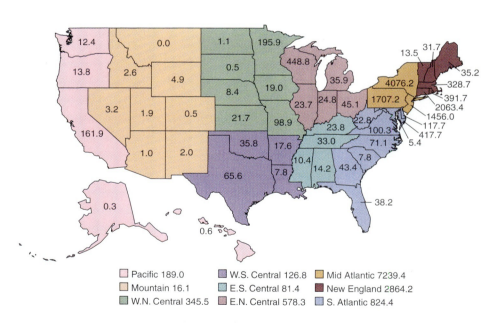

Average Number of Reported Cases of Lyme Disease per Year 1990–1999
Figure 22.14

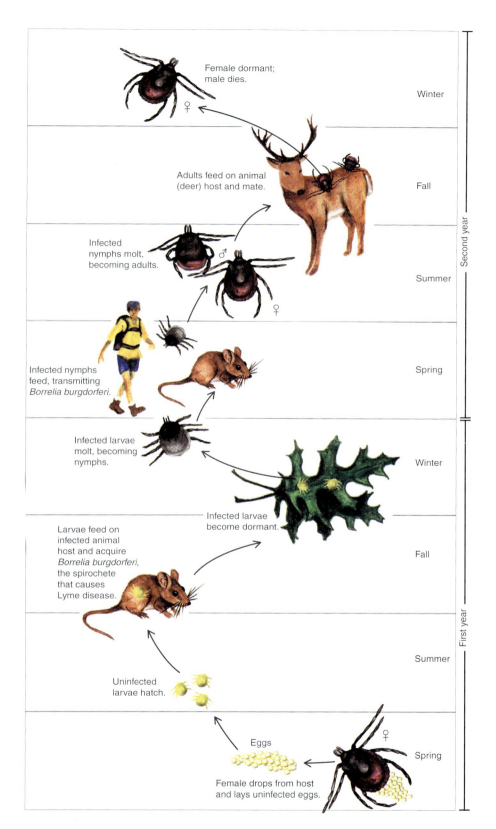

Life Cycle of the Black-Legged Tick, *Ixodes scapularis*, the Principal Vector of *Borrelia burgdorferi*, Cause of Lyme Disease
Figure 22.16

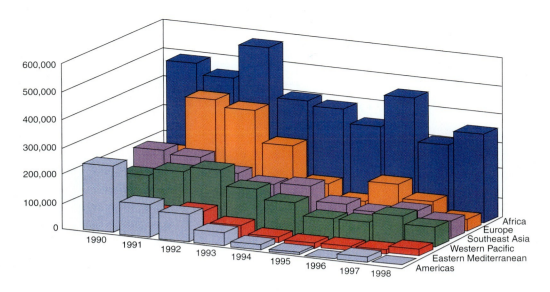

Reported Incidence of Measles in Different Regions of the World, 1990–1998
Figure 22.22

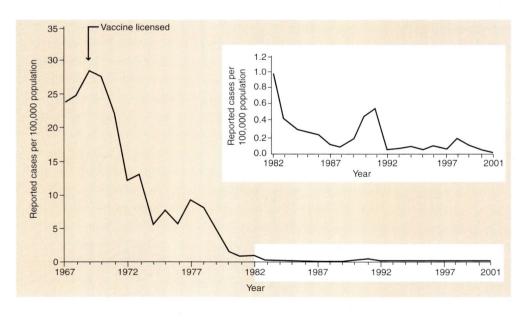

Reported Cases of German Measles (Rubella), United States, 1967–2001
Figure 22.24

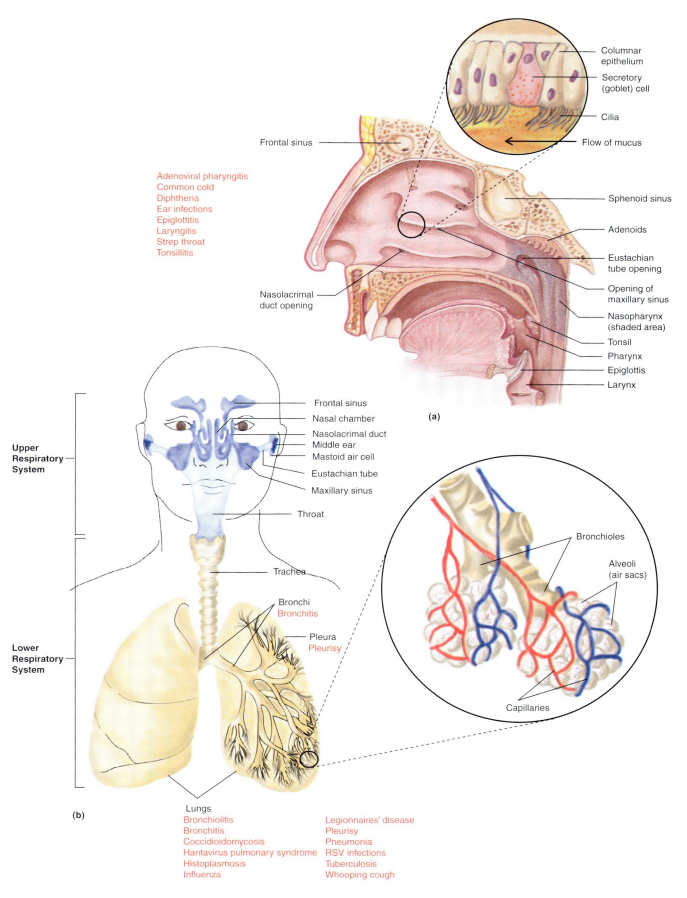

Anatomy and Infections of the Respiratory System
Figure 23.1

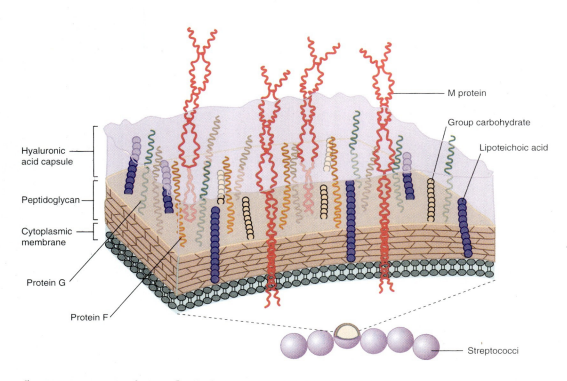

Components of the Cell Envelope of *Streptococcus pyogenes*
Figure 23.3

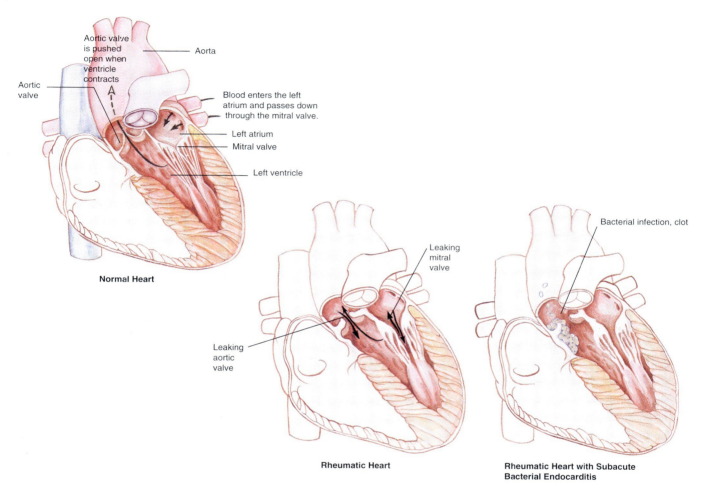

Rheumatic Heart Disease
Figure 23.4

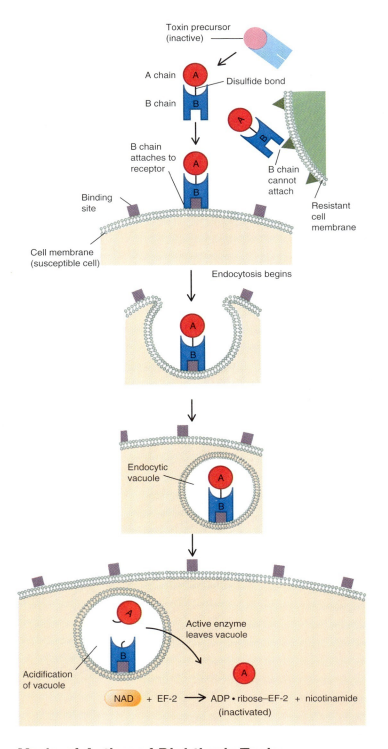

Mode of Action of Diphtheria Toxin
Figure 23.6

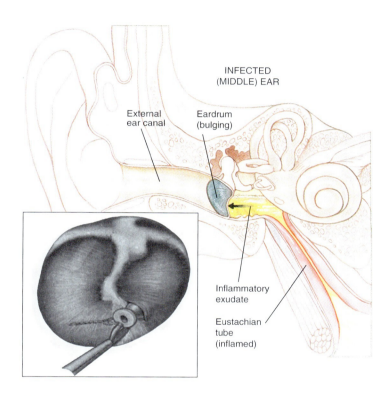

Otitis Media
Figure 23.7
Courtesy Smith & Nephew Ent.

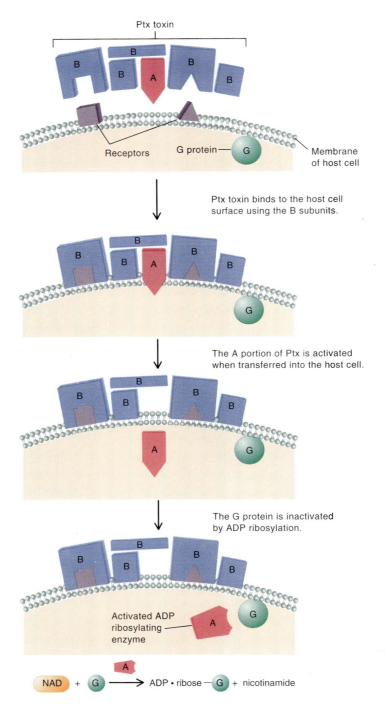

Mode of Action of Pertussis Toxin
Figure 23.14

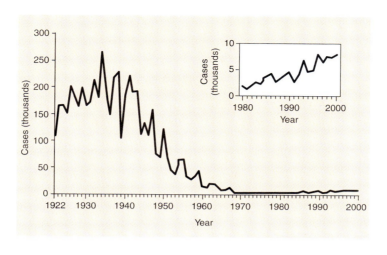

Number of Reported Pertussis Cases, by Year, United States, 1922–2000
Figure 23.15

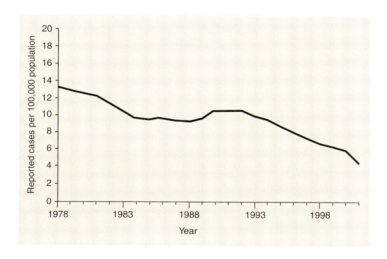

Incidence of Tuberculosis, United States, 1978–2001
Figure 23.16

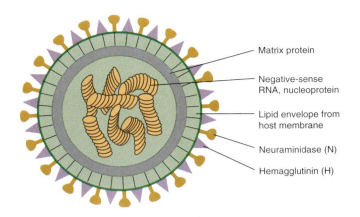

Diagrammatic Representation of Influenza Virus
Figure 23.21

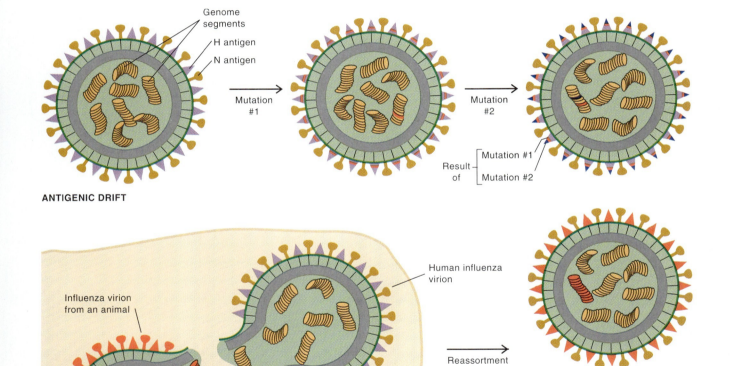

Influenza Virus: Antigenic Drift and Antigenic Shift
Figure 23.22

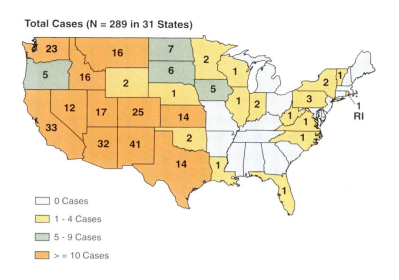

Hantavirus Pulmonary Syndrome Cases, United States, as of January 30, 2002
Figure 23.23

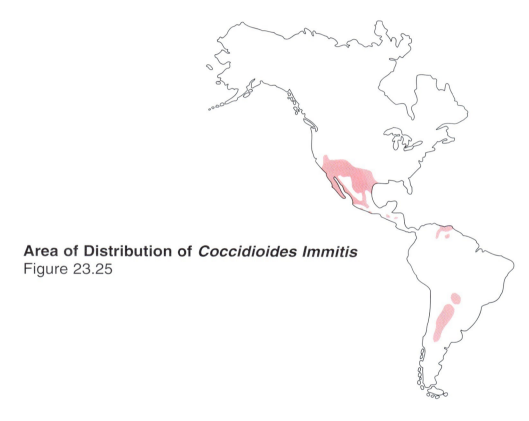

Area of Distribution of *Coccidioides Immitis*
Figure 23.25

Geographic Distribution of *Histoplasma capsulatum* in the United States as Revealed by Positive Skin Tests
Figure 23.27

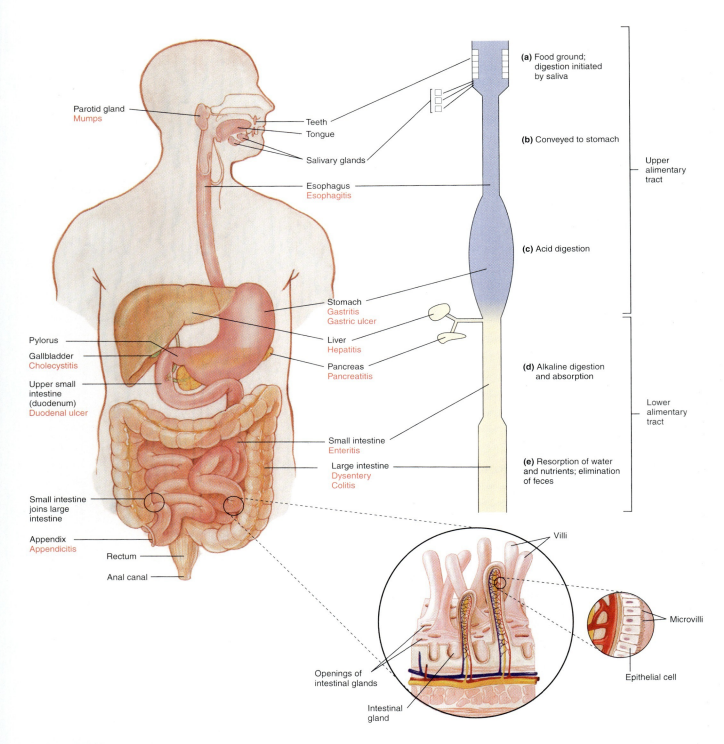

The Alimentary System
Figure 24.1

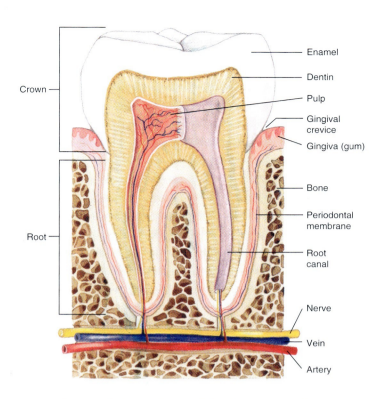

Structure of a Tooth and Its Surrounding Tissues
Figure 24.2

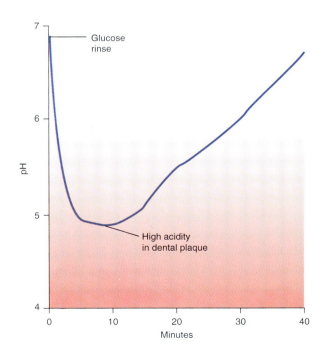

Increase in Acidity in Cariogenic Dental Plaque After Rinsing the Mouth with a Glucose Solution
Figure 24.4

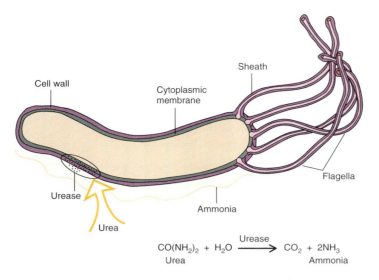

Helicobacter pylori
Figure 24.7

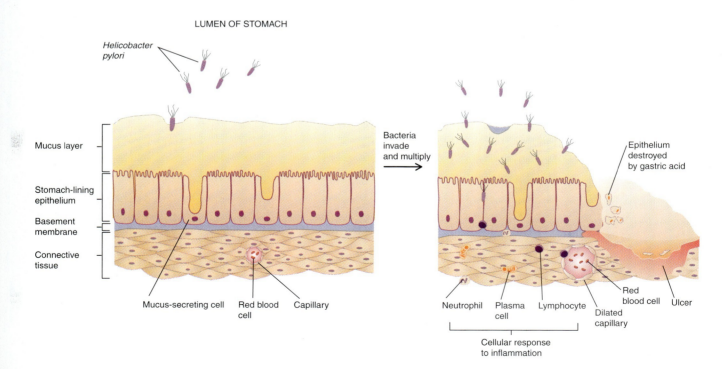

Gastric Ulcer Formation Associated with *Helicobacter pylori* Infection
Figure 24.8

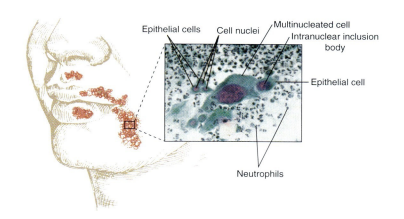

Herpes Simplex Labialis, Also Know As Cold Sores or Fever Blisters
Figure 24.9
© Frederick C. Skvara, M.D.

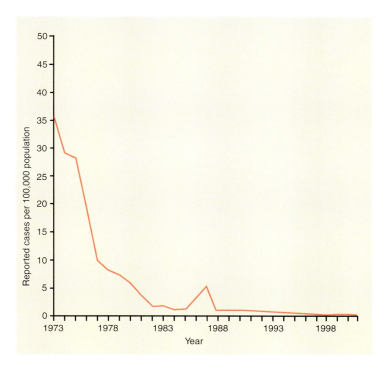

Reported Cases of Mumps, United States, 1973–2001
Figure 24.11

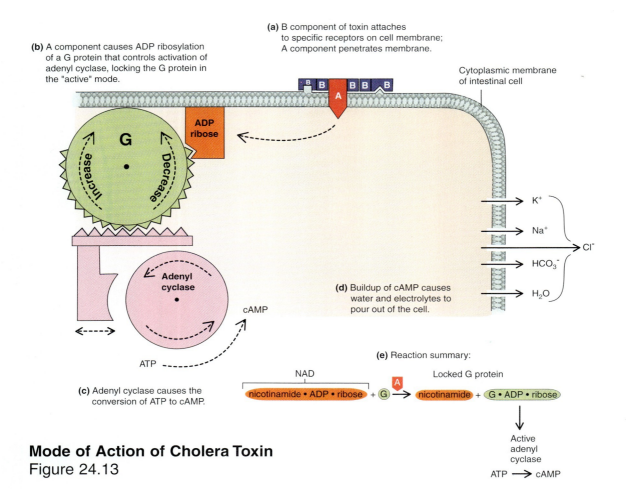

Mode of Action of Cholera Toxin
Figure 24.13

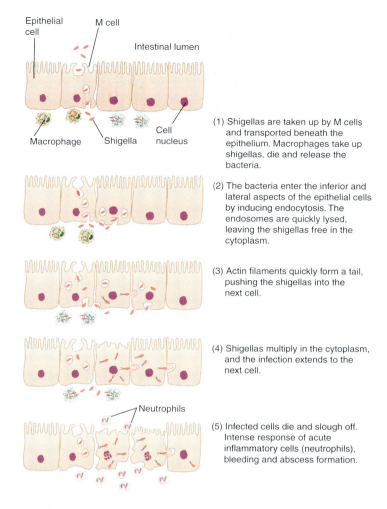

Pathogenesis of Shigellosis
Figure 24.14

Reported Cases of Hepatitis A, United States, 2001
Figure 24.18

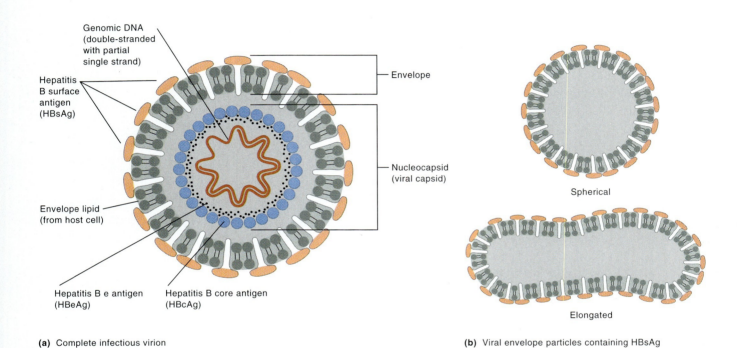

(a) Complete infectious virion

(b) Viral envelope particles containing HBsAg

Hepatitis B Virus Components Found in the Blood of Infected Individuals
Figure 24.19

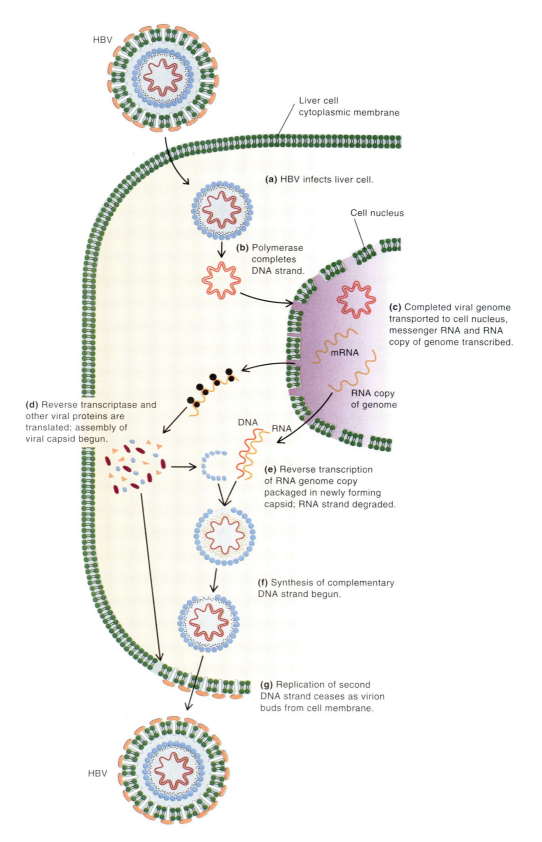

Replication of Hepatitis B Virus
Figure 24.20

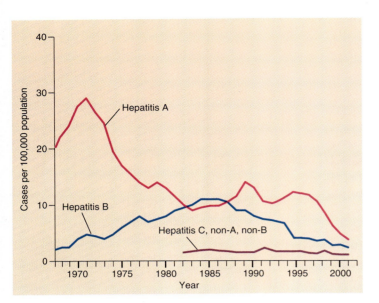

Incidence of Viral Hepatitis in the United States
Figure 24.21

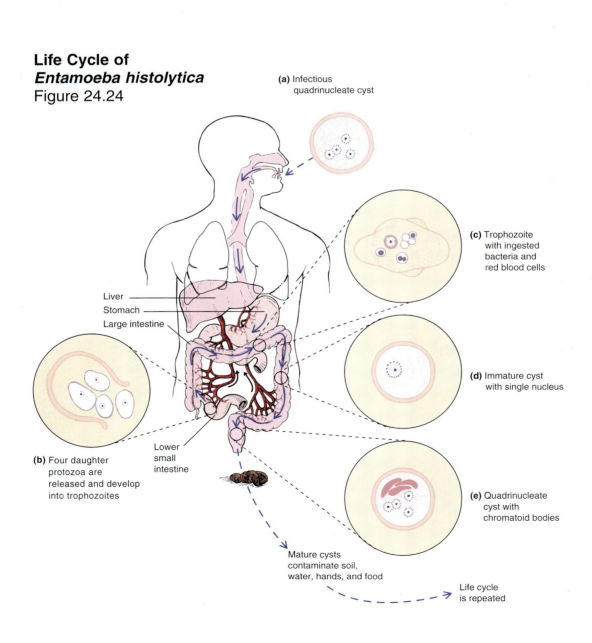

Life Cycle of *Entamoeba histolytica*
Figure 24.24

(a) Infectious quadrinucleate cyst

(b) Four daughter protozoa are released and develop into trophozoites

(c) Trophozoite with ingested bacteria and red blood cells

(d) Immature cyst with single nucleus

(e) Quadrinucleate cyst with chromatoid bodies

Mature cysts contaminate soil, water, hands, and food

Life cycle is repeated

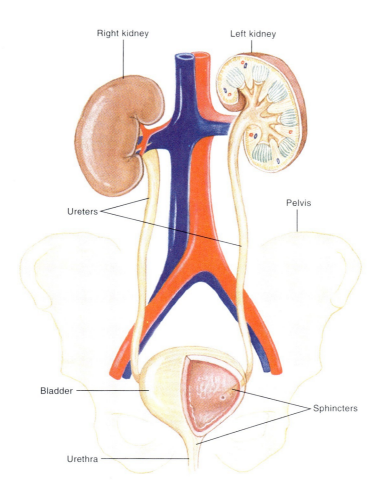

Anatomy of the Urinary System
Figure 25.1

Anatomy of the Genital System
Figure 25.2

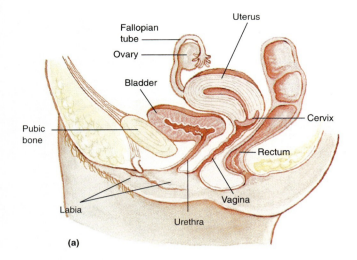

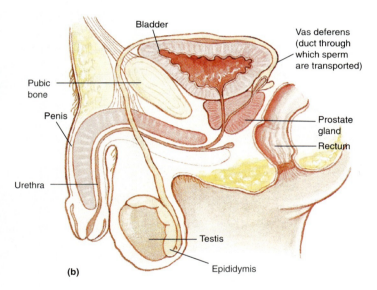

Staphylococcal Toxic Shock, United States, 1979–1996

Figure 25.6

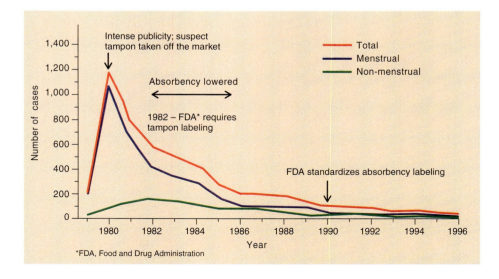

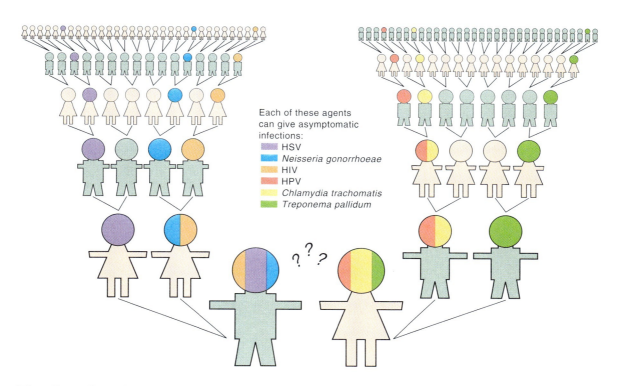

The Possible Risk of Acquiring a Sexually Transmitted Disease in Two Individuals Contemplating Unprotected Sexual Intercourse
Figure 25.7

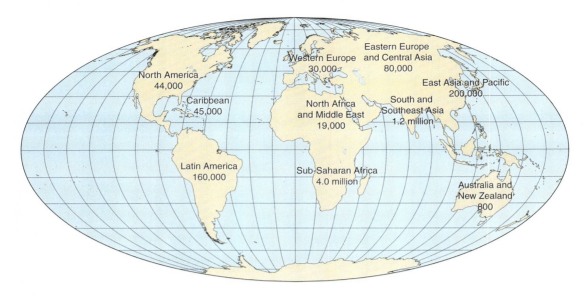

The AIDS Pandemic Continues Without Letup
Figure 25.20

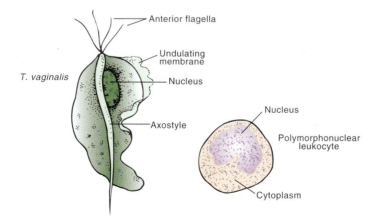

***Trichomonas vaginalis*, a Common Sexually Transmitted Cause of Vaginitis**
Figure 25.21

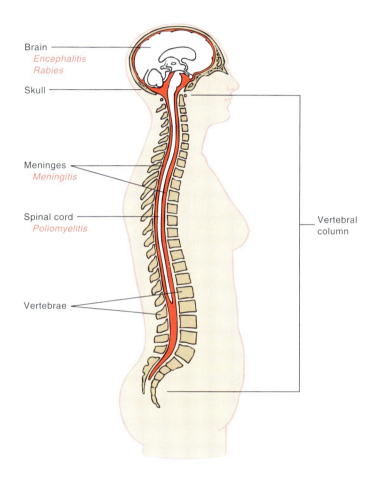

The Central Nervous System
Figure 26.1

Cerebrospinal Fluid
Figure 26.2

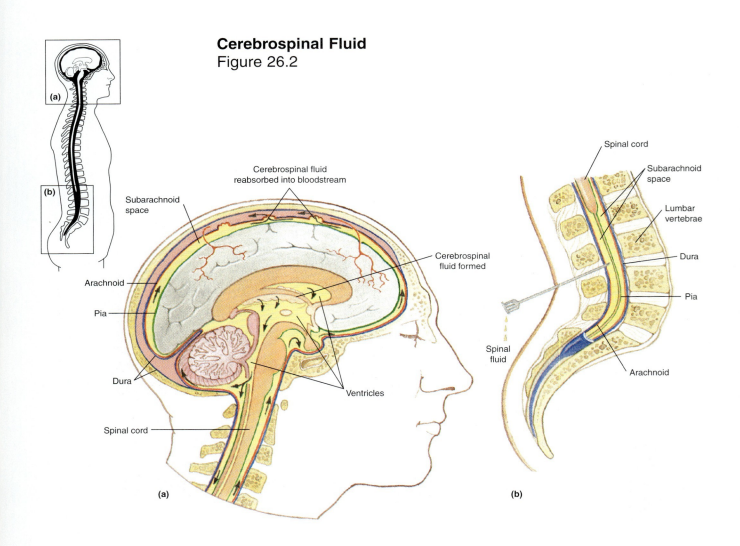

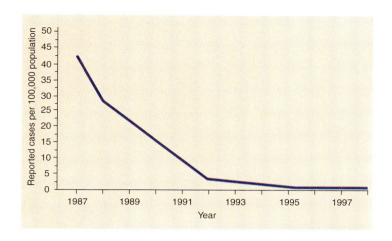

Rate of Serious *Haemophilus Influenzae* Disease per 100,000 Children Less than Age Five, United States, 1987 through 1998
Figure 26.3

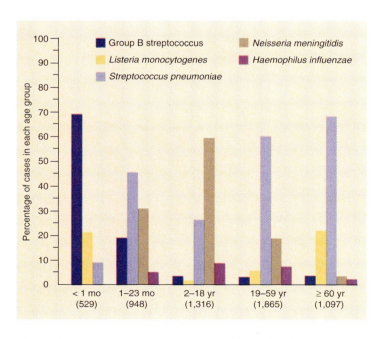

The Five Leading Causes of Meningitis in the United States, 1995
Figure 26.5

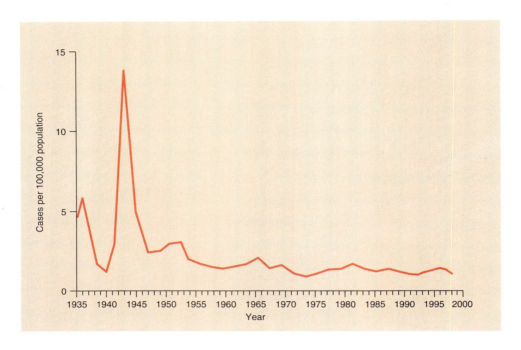

Meningococcal Disease in the United States, 1935 to 1998
Figure 26.8

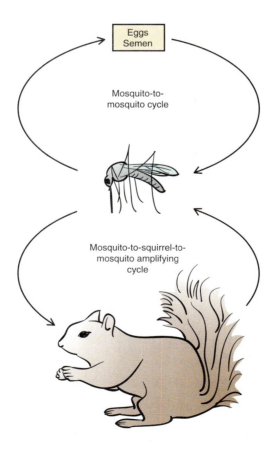

LaCrosse Encephalitis Virus, Natural Cycles
Figure 26.13

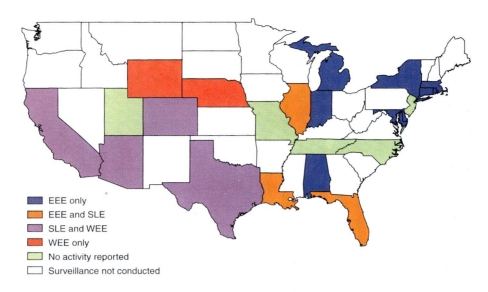

Distribution of Encephalitis-Causing Arboviruses
Figure 26.14

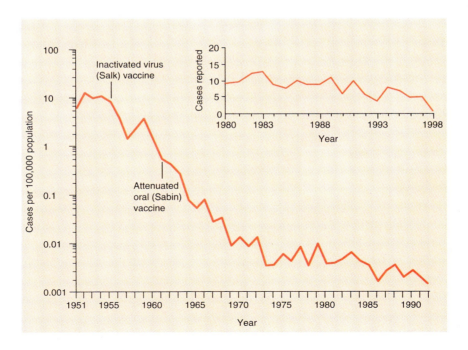

Incidence of Poliomyelitis in the United States, 1951 to 1998
Figure 26.18

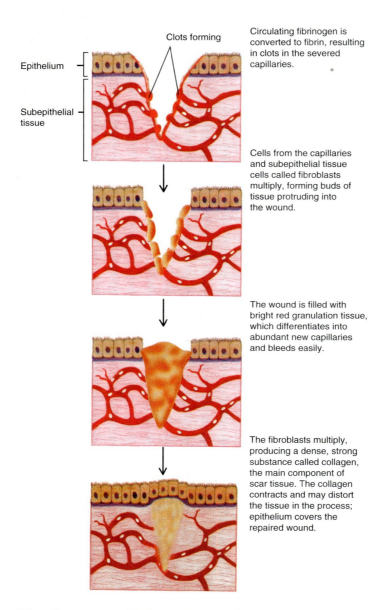

The Process of Wound Repair
Figure 27.1

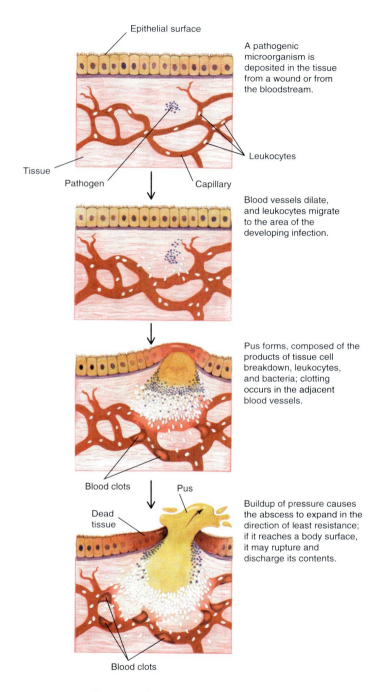

Abscess Formation
Figure 27.2

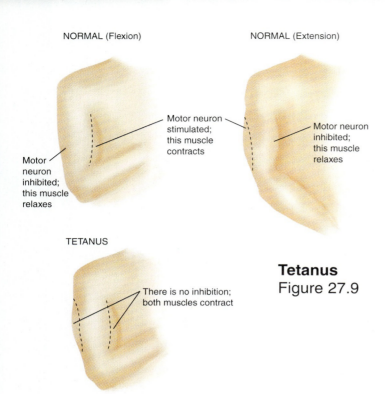

Tetanus
Figure 27.9

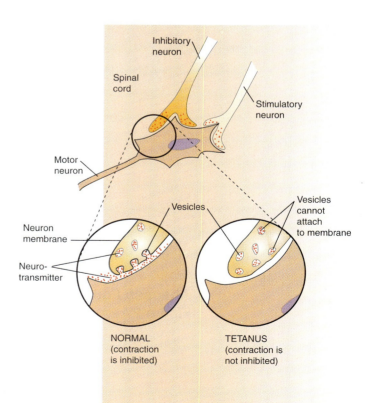

Inhibitory Neuron Function
Figure 27.10

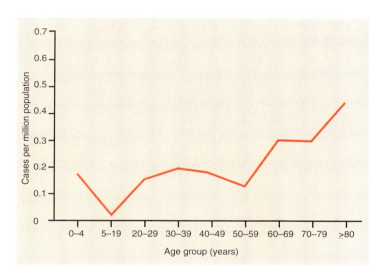

Average Annual Incidence of Tetanus for Different Age Groups, United States, 1995 to 1997
Figure 27.11

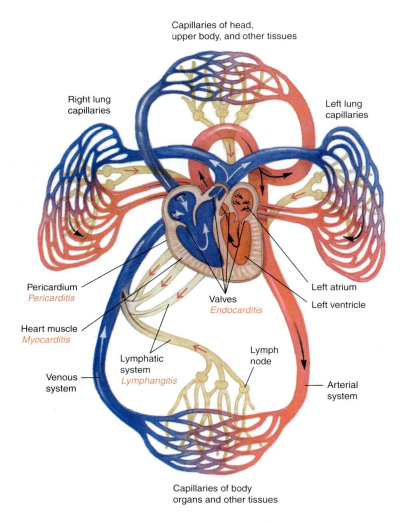

The Blood and Lymphatic Systems
Figure 28.1

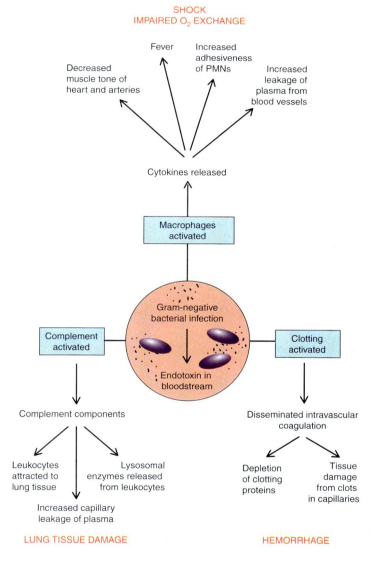

Events in Gram-Negative Septicemia
Figure 28.3

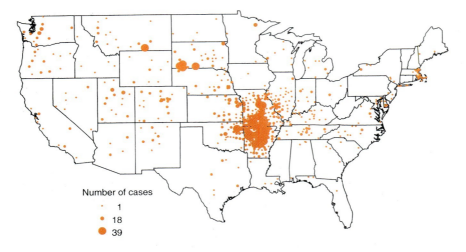

Reported Cases of Tularemia, United States, 1990–2000
Figure 28.5

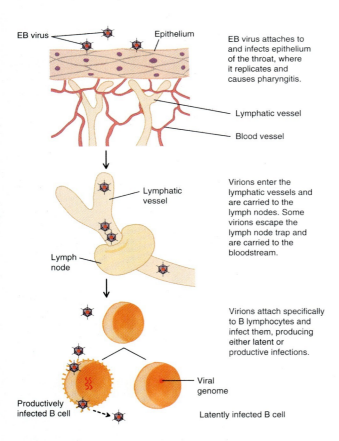
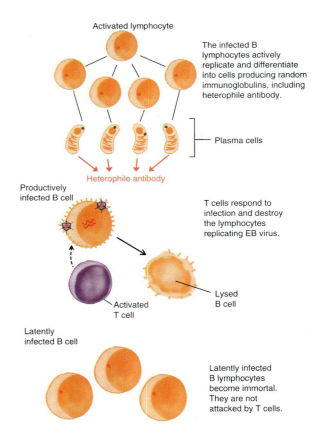

Pathogenesis of Infectious Mononucleosis
Figure 28.8

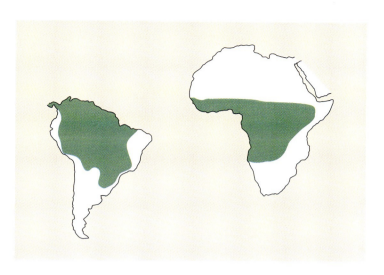

Distribution of Yellow Fever
Figure 28.10

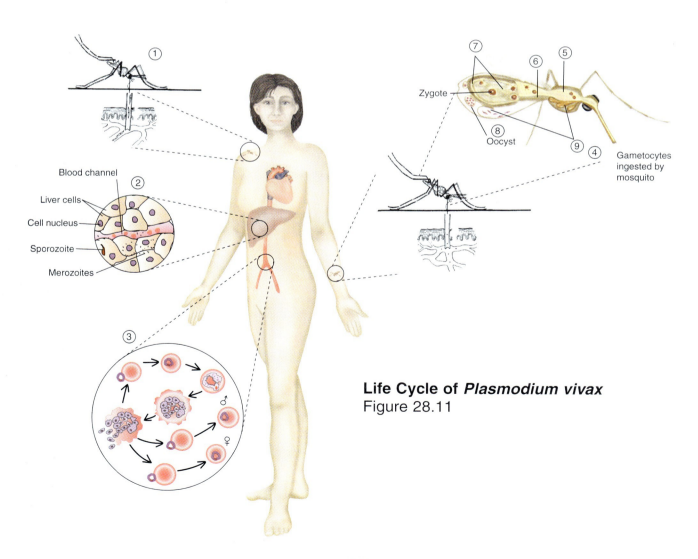

Life Cycle of *Plasmodium vivax*
Figure 28.11

231

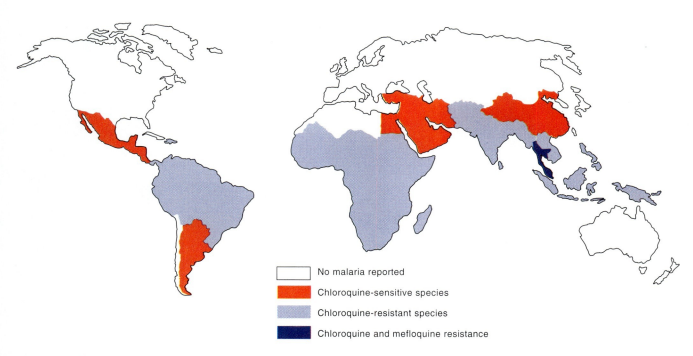

Distribution of Malaria in 1996
Figure 28.12

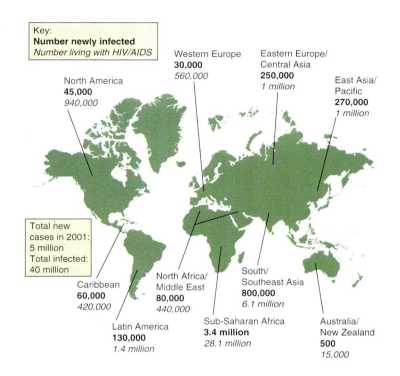

The HIV/AIDS Epidemic Two Decades After the Onset
Figure 29.1

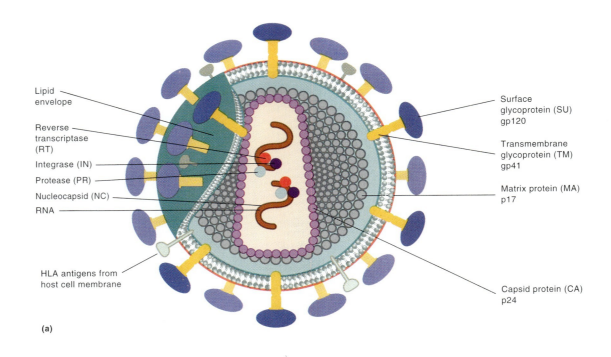
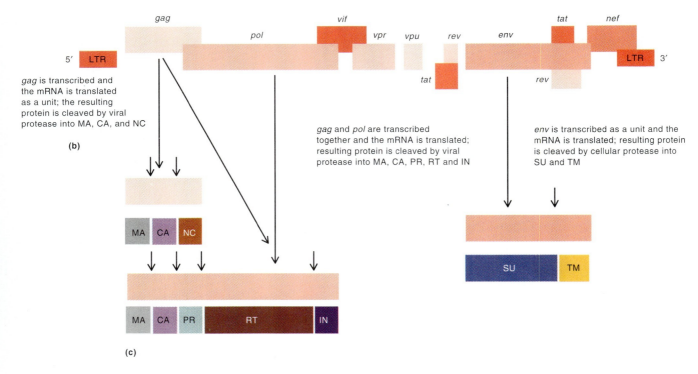

Human Immunodeficiency Virus, Type 1 (HIV-1)
Figure 29.3

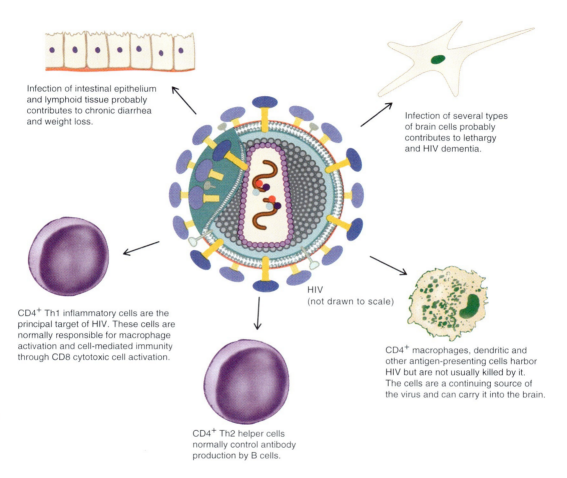

Some of HIV's Cellular Targets
Figure 29.4

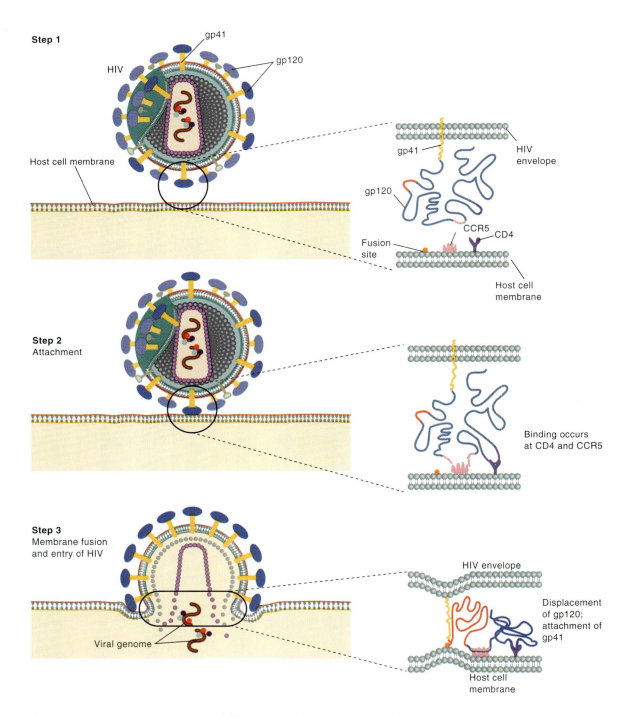

Attachment and Entry of HIV into a Host Cell, Schematic Representation
Figure 29.5

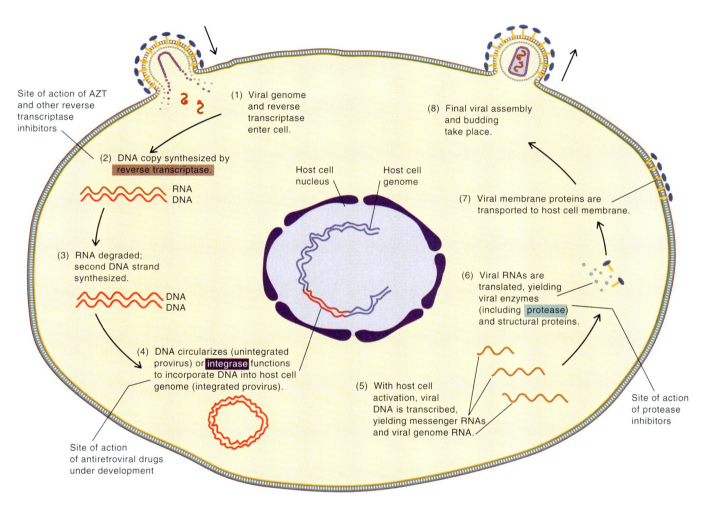

The Steps in HIV Replication
Figure 29.6

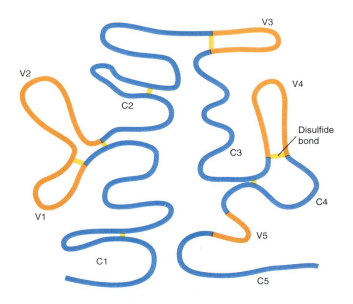

Diagram of the SU Glycoprotein
Figure 29.7

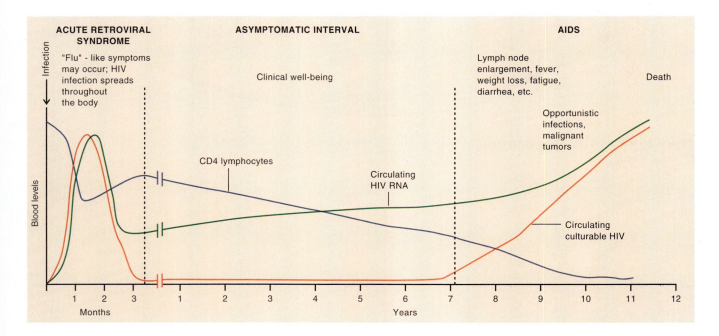

Natural History of HIV Disease
Figure 29.8

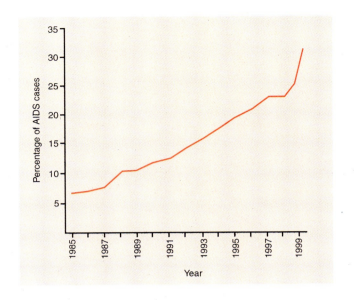

In the United States, a Steadily Rising Percentage of AIDS Cases Occur in Women
Figure 29.9

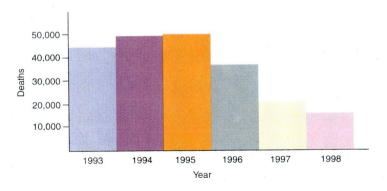

Estimated Deaths Due to AIDS, United States, 1993 to 1998
Figure 29.10

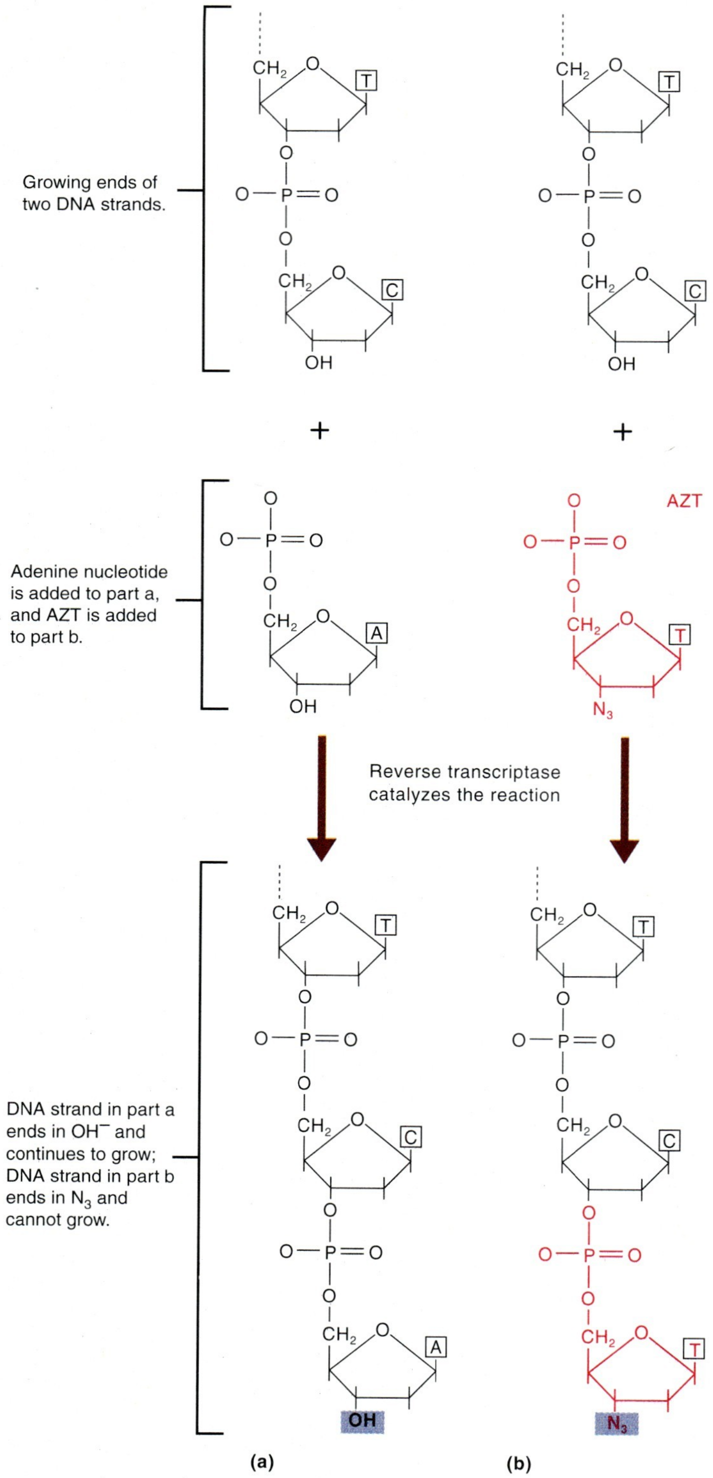

Mode of Action of Zidovudine (AZT)
Figure 29.11

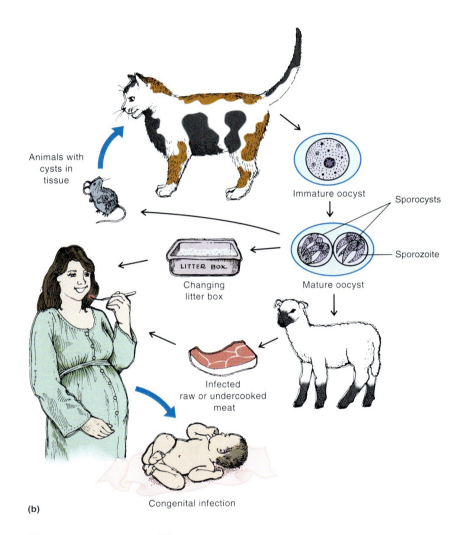

Toxoplasma gondii
Figure 29.14

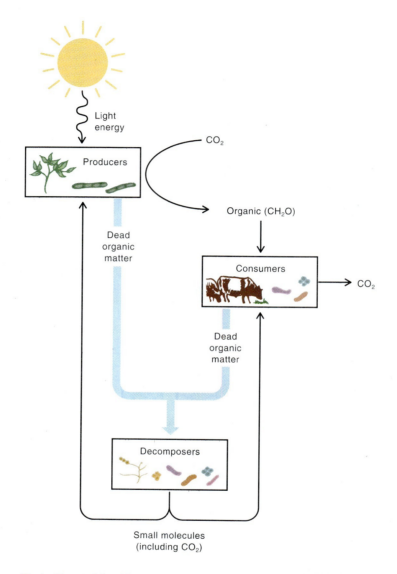

Relationship Between Producers, Consumers, and Decomposers in an Ecosystem
Figure 30.2

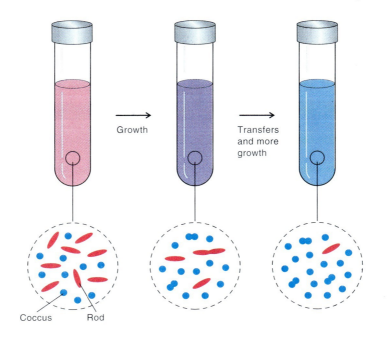

Competition Between Two Bacteria
Figure 30.3

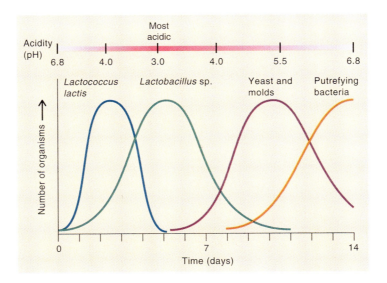

Growth of Microbial Populations in Unpasteurized Raw Milk at Room Temperature
Figure 30.4

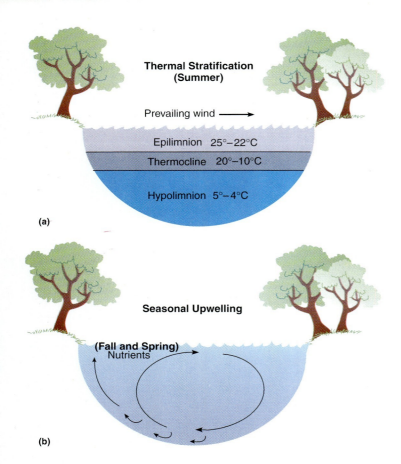

Thermal Stratification of a Lake
Figure 30.7

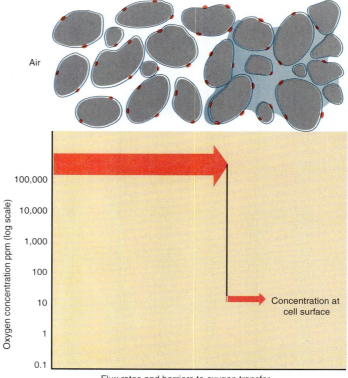

Texture of Soil
Figure 30.8

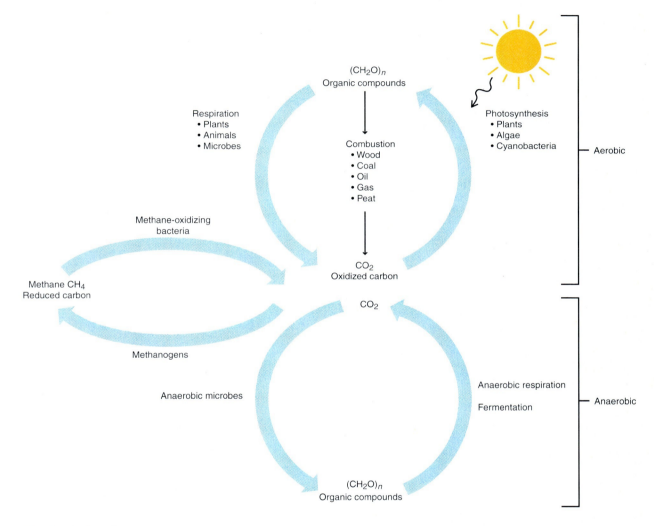

Carbon Cycle
Figure 30.9

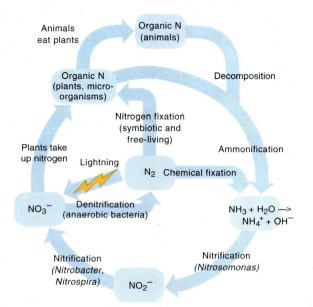

Nitrogen Cycle
Figure 30.11

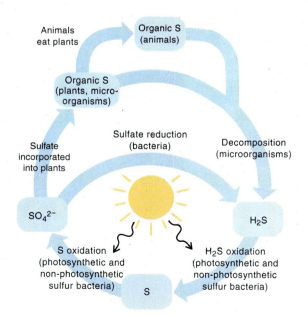

Sulfur Cycle
Figure 30.12

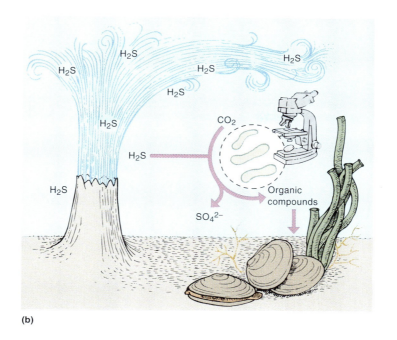

Hydrothermal Vent Community
Figure 30.13

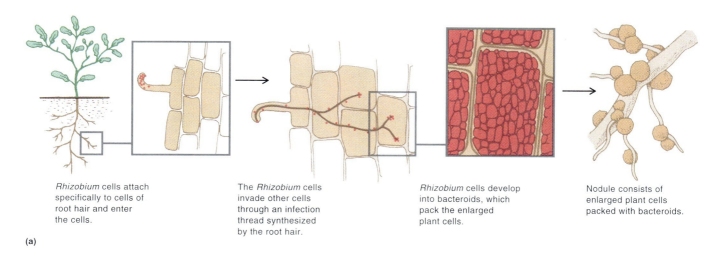

Rhizobium cells attach specifically to cells of root hair and enter the cells.

The *Rhizobium* cells invade other cells through an infection thread synthesized by the root hair.

Rhizobium cells develop into bacteroids, which pack the enlarged plant cells.

Nodule consists of enlarged plant cells packed with bacteroids.

(a)

Symbiotic Nitrogen Fixation
Figure 30.14

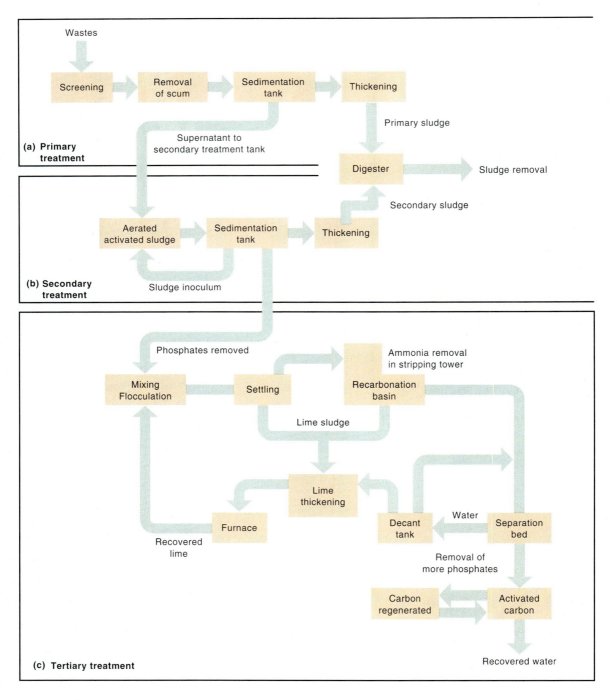

Municipal Sewage Treatment
Figure 31.1

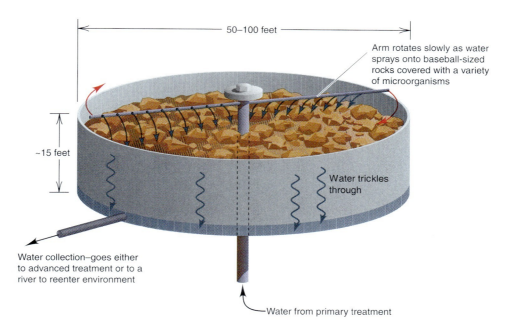

Trickling Filter
Figure 31.2

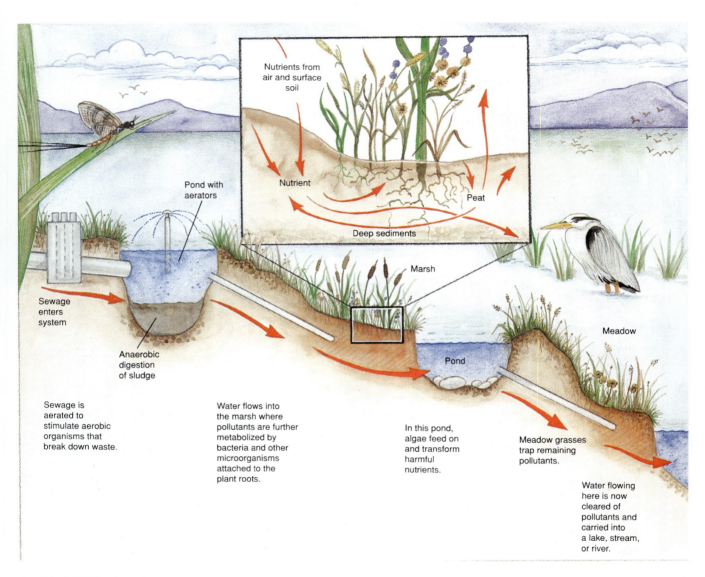

Artificial Wetland
Figure 31.3

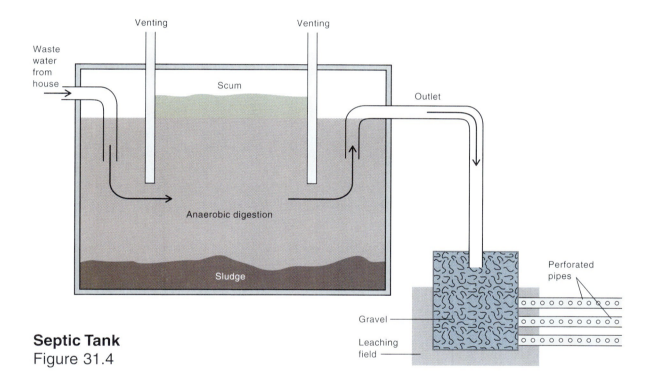

Septic Tank
Figure 31.4

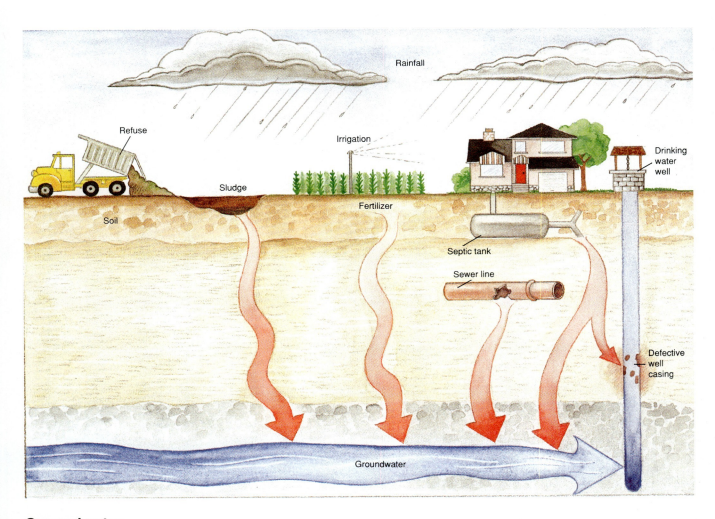

Groundwater
Figure 31.5

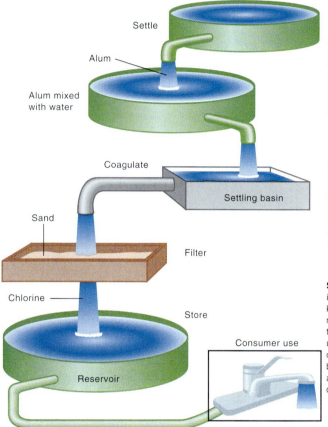

Steps in the Treatment of Metropolitan Water Supplies
Figure 31.6

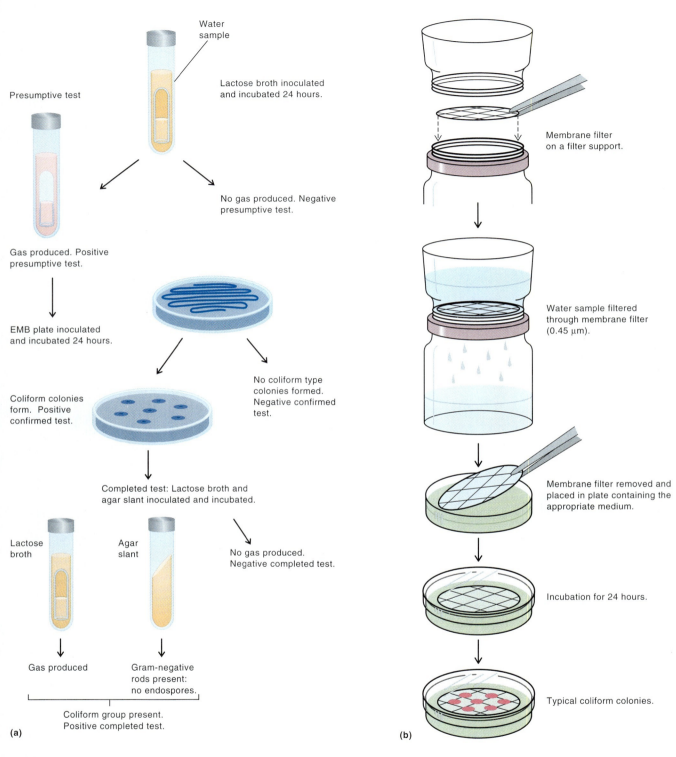

Methods Used for Testing Water
Figure 31.7

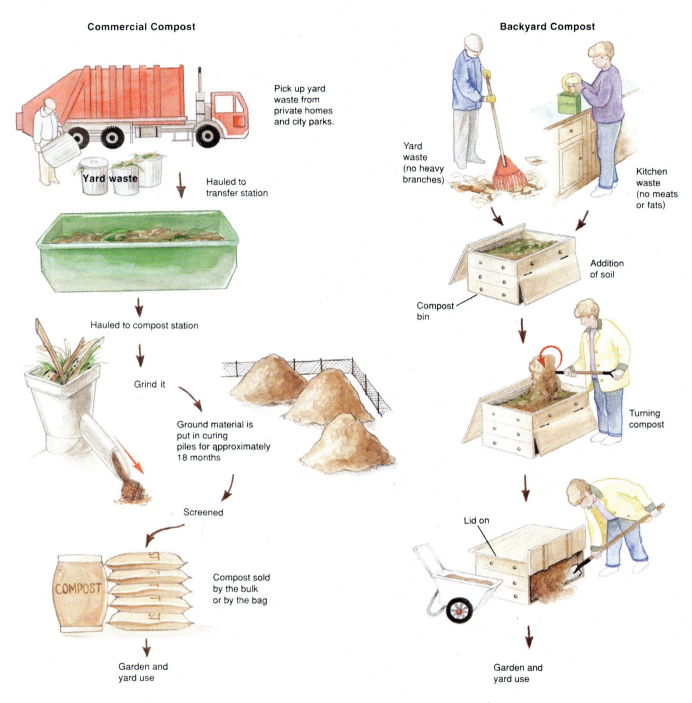

Industrial and Backyard Composting
Figure 31.9

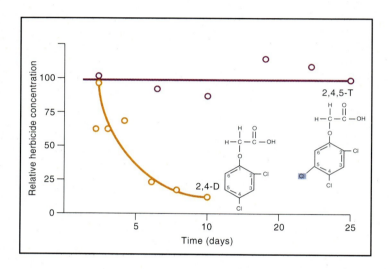

Comparison of the Rates of Disappearance of Two Structurally Related Herbicides, 2,4-D, and 2,4,5-T
Figure 31.10

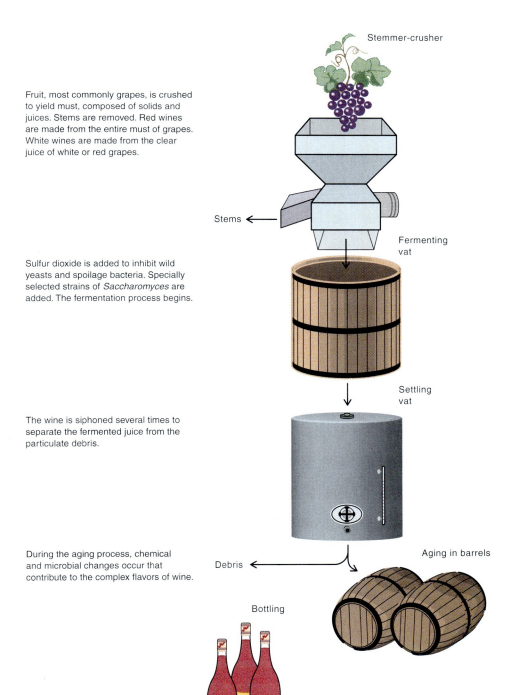

Commercial Production of Wine
Figure 32.4

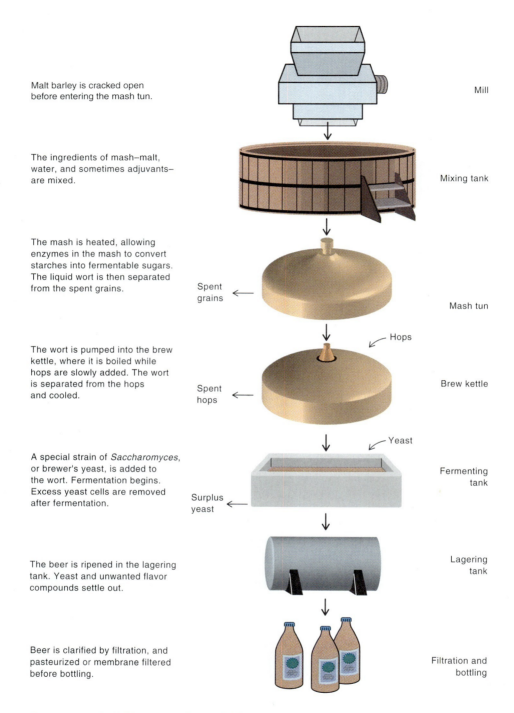

Commercial Production of Beer
Figure 32.5